DON EMILIO

DON EMILIO

Donald E López

ATHENA PRESS
LONDON

DON EMILIO
Copyright © Donald E López 2005

All Rights Reserved

No part of this book may be reproduced in any form
by photocopying or by any electronic or mechanical means,
including information storage or retrieval systems,
without permission in writing from both the copyright
owner and the publisher of this book.

ISBN 1 84401 518 1

First Published 2005
ATHENA PRESS
Queen's House, 2 Holly Road
Twickenham TW1 4EG
United Kingdom

Printed for Athena Press

Author's Note

This book was essentially written for my family, to acquaint them with my earlier life, as most of my business activities were known to them already. However, in order to satisfy those friends of mine who were disappointed to find that I had not continued my autobiography to cover the successful business period that was to follow, I will mention some of those activities and events here.

After the events narrated in this book, I took over several companies, one making high quality garden furniture from renewable hardwood, and another a well known and established luxury lighting company. I designed, patented and manufactured such well known luggage as Kingfisher, Crown, Revellation and finally, Antler Luggage, the best and most expensive British-made luggage.

At one time I had the largest RF welding company in the UK, which manufactured a wide range of PVC products. I designed and patented the earliest APM (Alternating Pressure Mattress) – now commonly known in the NHS as "ripple beds". For its manufacture, I built an entirely new factory.

I designed and manufactured a great array of products for Marks and Spencer. I also produced tens of thousands of cool bags for Thermos. I set up a tool hire company under the name of Lark Hire, a subsidiary of Olivelark, which I had formed in the early 1970s. In all, I took out forty-seven patents, covering a wide range of products and manufacturing methods.

I also designed and manufactured "on the hanger travelpacks – the first in this country. These were made very popular by Selfridges in their weekly *Sunday Express* offers, which carried on for many years. Shell and BP also made these an essential piece of luggage for their executives, as they were designed to be taken into aeroplane cabins, thus avoiding lengthy waits in the baggage reclaim area, and Patrick McNee in *The Avengers* never failed to illustrate its qualities during a fast getaway scene!

I modified several of my large travelpacks for Prince Philip, so that each one contained the complete official uniform for each of the Armed Forces; this made life much easier for his valet!

I made and designed for Burberry, Aquascutum, Zandra Rhodes and many more. I became a member of the Institute of Patentees and Inventors, and met many famous inventors, as well as Sir Frank Whittle's assistant. I became the 441st Master of the Incorporation of Weavers, Fullers and Shearmen in 1989 – a guild which was granted its Royal Charter in January 1559 and not only survived the Reformation, but also the bombing of the beautiful Tucker's Hall in the Second World War.

Regrettably, none of my daughters wished to take over the business and the five factories, employing several hundred personnel. I was left with few alternatives, and consequently, I decided to sell up!

Foreword: La familia

Don Emilio was born into a wealthy family on 25th January 1926. He was the third child of Amilia López García and Guillermo López Echeverría de Peralta, who was affectionately known by the family as Willy. There was nothing particularly special about Don Emilio, other than that from the day he was born, he was to be called Don Emilio whenever addressed by the servants, of whom there were many, and by anyone else outside the family.

The house in Plaza Vieja was of a reasonable size, although not large in comparison to some of the other properties belonging to Don Emilio's grandparents, Señor Don Guillermo López Rull and La Marquesa Ana Echeverría de Patrullo. Sadly, Don's grandmother Ana had died seven years before Don Emilio was born. She was buried in the family vault on the outskirts of Almería in southern Spain; an area now generally known as the Costa del Sol.

Don Emilio lived for a brief time with his brother and sister at Plaza Vieja, which was virtually at the centre of the city and close to the telephone exchange, where their father was the managing director. It was at the exchange that Guillermo had met Don Emilio's mother, Amilia García, who at the time was working as a telephone operator. After a short and stormy period together, and in spite of bitter family objections, Guillermo and Amilia were married. Sadly, Guillermo's family were of the opinion that Amilia was unsuitable for their son, having come from a working-class background. But Amilia was very beautiful and Guillermo had become immediately smitten with her, beyond recourse.

When Manolo was born, the fourth youngest child of Guillermo and Amilia, the family moved to El Depósito de Agua, which was a ranch-style property off the Barrio Alto, a suburb of Almería. The house was large and far more suitable for the enlarged family, however, a good proportion of the property was devoted to offices for the water company. This was because the house was sited in the middle of a series of reservoirs, deep wells and housing for pumping machinery. It was not exactly beautiful by any stretch of the imagination, but it was a positive money spinner for Don Emilio's grandfather, Don Guillermo, as water in that part of Spain was valued like liquid gold... and still is!

Again, the stay in the property at El Depósito de Agua was short-lived and the family moved once again when Guillermo Junior was born, this time to La Pipa. This house was considered to be 'the jewel in the crown', so to speak, by the family as it was a very beautiful house, in total contrast to El Depósito. It was surrounded on three sides by high walls; between the house and the outer walls were trees, strategically placed to give maximum seclusion, and there were formal gardens extending up to the terraces.

The roof tiles were curved and bottle green in colour, and the house itself was white. Although from the front the house was hidden from general view, the rear overlooked two terraces, each with balustrades about forty feet wide. There were steps leading from the upper to the lower terrace, which ran alongside the railway line. It was possible to look out of the ground floor windows and see the distant villages of Molinos, Chocillas and Cuatro Caminos in the views surrounding the property, as La Pipa had been built on the commanding position of a small hillock.

Beyond the two terraces and the stone balustrades, there were tennis courts and past these, a rose garden that went as far as the fence of the main railway line, which was the connecting line between the cities of Madrid, Granada and Almería and had been built by Señor Don Guillermo López Rull, Don Emilio's grandfather.

La Pipa was spacious and graceful. The family crest was embossed over the elaborate front entrance, which also consisted of a pair of solid Spanish oak doors, each having wrought-iron grill windows and hung in the centre of a large arch. Each ground floor window was placed between pillars, under more arches, and the upper floor windows were placed between half-size pillars. The front doors led to an octagonal hall, with a white and blue mosaic tiled floor. From here, through an imposing, rich, dark oak door, you entered a long corridor. On one side, above the oak panelling on the walls, there were tall, mullioned windows with coloured glass. On the other side, large oil paintings were hanging above the panelling, a good proportion of which were the portraits of the family, depicting the more important members who had helped to shape the vivid and exciting family history.

La Pipa was, and still is, a tourist attraction and was but a small part of the many estates owned by Señor Don Guillermo. All were within easy reach of the others, La Torre was a huge house with a tower, being its predominant feature; not far away was Villa Anita and El Cortijo de López. The main town house was in the middle of the Plaza del Conde Ofalia and was used for winter functions and was also the control centre of the many family businesses. During the Civil War, the house became the headquarters of the *Falanje* (Fascist party), and later it became the judicial court house.

However, La Torre was by far the most luxurious of the properties. It had a gigantic reservoir for irrigation, a swimming pool and a glorious formal garden. A short distance from the house was a home farm, consisting of approximately 2000 acres of mixed farming. From the top of the house's tower, there were magnificent views of the Sierra Alhamilla and the port of Almería. The sitting room had French windows, which led to a patio, and that in turn to a flight of steps leading to a sunken garden with rose beds and shady arbours. The quality of the house could be judged, for example, by looking at the main bathroom, which was more like a temple, with its pillars and a bath made from pink alabaster in the shape of an oyster shell.

La Pipa, however, had become the home for the extended family, which at times would include grandfather Don Guillermo, if he was not residing in another of his many properties. Many servants and footmen ensured that the house functioned smoothly and efficiently, despite the many comings and goings of its ever-mobile residents.

During the week, grandfather Don Guillermo lived at the large town house, which had been designed for him by his brother, Enrique. It faced the centre of the gardens in Plaza Conde de Ofalia, in Almería. It was from this house that Don Guillermo controlled his many businesses and also hosted many lavish parties. However, his most renowned gatherings were held at La Pipa and for such occasions, he would send his footmen and the family coaches to Lhardy in Madrid for the necessary provisions. The more bulky items needed were sent by train, which would deliver by making a special stop at the platform built at the lower garden boundary of La Pipa, in order for it to be able to unload. The passengers on the train had no choice in the matter... they just had to wait! Almería would soon know if a grand party at La Pipa was imminent.

From June to September, some of the family would go to La Hoya del Molinero, a big rambling farmhouse (known as a *hacienda*), near Olula in the valley formed by the Sierra de las Estancias and the Sierra de los Fibrales. This was the old home of Pedro López García and his wife, Concepción Rull Rueda. Pedro was a master mason and the couple owned the marble quarries in and around Cortijo de los Navarros. To get to La Hoya de Molineros was like going on an expedition! The only way to get to it was by mule train. The string of mules not only carried the family and servants, but also all of the provisions that would be required for a long stay. On one such occasion, the family even transported a grand piano on the back of a mule, so that one of the children or aunts could continue with their piano lessons!

The mule train was always accompanied by the family guards, who were there to protect them from brigands and outlaws. The guards wore uniforms with leather doublets, which displayed the family coat-of-arms. A small number of the staff remained permanently at the *hacienda* – these would have been the most loyal and trustworthy who could therefore be left to their own devices during the long, cold winter months.

La Hoya de Molineros was not only a wonderful country residence, but also served as both a holiday retreat and a power-house for the esparto grass business. Esparto grass was considered at that time to be the most important ingredient for the manufacture of top quality paper and was business for Don Guillermo. Business people and office colleagues of his would very often be invited to join the shooting parties and go hunting at La Hoya de Molineros. Sometimes they would stay for several days at a time, especially if the weather proved particularly bad.

★

Grandfather Don Guillermo was the eldest of nine children and because of the influence exerted on him by his part-English and part-American father, he became an arch anglophile. His favourite of his own children was Don Emilio's own father, Willy, whom Don Guillermo considered, in his own opinion, to epitomise the typical English gentleman. So, for that reason, Willy was often referred to as 'Willy, El Inglés' (Willy, The English); a nickname that was to be inherited by Don Emilio on the early death of his father.

★

On 24th September 1995, an article appeared in a local newspaper of Almería, called *The Voice of Almería*, with the following details regarding the family.

THE LÓPEZ RULL... THE GENIUS OF THE SUCCESSFUL

The family originated from Olula del Río, the centre of the marble quarry industry. This family has, for the last century and up to the present, been responsible for shaping a large portion of the city of Almería.

A branch of the family, the Echeverrías, with roots in Navarra, were destined to create great wealth, not only for themselves, but also for the city. They brought considerable business experience and the necessary capital. They become the new landed gentry of nineteenth-century Almería. The city was largely fashioned by Señor Don López Rull, the first son of that courageous mother, who made no promises to her children so, therefore, they had no greater expectations than to follow their father into the family quarry business and remain there for the rest of their working lives.

Pedro López García and Concepción Rull Rueda married in Olula del Río in 1837. Both came from Olula. He worked as a master mason in the marble quarries. They had a daughter called

Carmen and three boys. Concepción, who saw her husband work from sunrise to sunset, decided that her children would have better prospects in life than her hard-working husband, so when her boys were about to begin their studies for the matriculation examinations, she took them to Almería to live and gave them the opportunity of making fresh careers for themselves and avoid being swallowed up by the quarries.

Concepción's decision was thoroughly approved of by her husband, who now had to work even harder to keep two homes going. On Saturdays when he finished work, he went to see his wife and children in Almería. On Monday mornings at the crack of dawn, he had to make the long and arduous return journey on horseback to the Olula quarries, as there was no other means of transport with there being no road, only a narrow and dangerous mule track. These were very hard years to cope with, but their exemplary marriage and good housekeeping made it all possible, in spite of having to live apart for so long. The children did them proud and Enrique became an architect, Guillermo and Pedro joined the armed services and Nicolas became a doctor.

Enrique López Rull became the leading architect in both Almería city and the province, from the mid-nineteenth century to the beginning of the twentieth century. Among his more famous works were: La Residencia de los Padres Domínicos in the Plaza Conde de Ofalia (now demolished); El Colegio de la Compañía de María; El Colegio del Paseo (originally the residence of Don Emilio Pérez Ibáñez and now El Casino); the shop in Plaza Conde de Ofalia and the corner on Calle Villa Espesa, which at one time belonged to his brother, Guillermo; Guillermo's house in the Paseo; La Casa de la Pena, in Plaza Emilio Pérez; the ancient monastery of Las Hermanitas de los Pobres; Convento de los Adoratrices; La Puerta de Purchena (this large house is now owned by the Quesada family); the façade of El Teatro Cervantes.

Enrique was also the Official Municipal Architect, the architect to Bishop Orbera and the architect to the bishop's diocese.

Guillermo López Rull settled for the armed services. His service history serves as a good example of nineteenth-century Spain and its revolutionary times. He swore the oath of allegiance as a cadet to His Majesty King Amadeo the first of Saboya. He passed out from the Estado Mayor Academy in 1872 as a commissioned officer. In 1874, he was named the Alférez Cadet of the Estado Mayor by order of the President of the First Republic, and on the 14th August 1875, he was promoted to lieutenant by command of King Alfonso XII.

In 1884, Guillermo came to Almería after deciding to retire from the army. Before his retirement, he had been very heavily involved in the Carlist Wars. Whilst based at Vitoria, he took part in the capture of Valmaseda on 29th January 1876. He was also in the fierce fighting for Monte Hernio. He disbanded the Carlist Army and returned to Navarra y Alava and subsequently went on to Vitoria and San Sebastián. Guillermo became a captain as a result of his meritorious service in the Carlist Wars.

On 4th July 1877, Guillermo married Ana Echeverría Patrullo in Pamplona cathedral. She was the eldest of five children and had been born in New York, so was a citizen of the United States of America. She was the daughter of Manuel Echeverría, a well-known Carlist supporter. Manuel had been exiled to France, where he educated his daughters at the Sacre Coeur, Paris. He owned a bank at Zapatería Street, Pamplona and a shipping business, which shipped to and from New York.

The Echeverrías were a very wealthy Navarra family, who on finding out that their daughter wished to marry an Isabelino military man who had fought against her father in the Carlist Wars, immediately opposed the marriage and put Ana into a convent. But Ana stuck to her guns and her father had no alternative but to give in. However, it was not until the actual day of the wedding that she was removed from the convent. Not a single member of Ana's family attended her wedding. Ana was 20 years old and her new husband, Guillermo, was 25 years old.

The newly wedded couple moved to Almería and on the birth of their first daughter, Concha, the two families became reconciled. Guillermo left for the war in Cuba in 1878 and when he returned, he continued his military career in Navarra and San Sebastián. When he retired in 1888 as Commander in Chief, Ana had given birth, by then, to ten children, but only six survived. The other four had died when they were very young.

In 1888 on the death of her father, Ana inherited a huge fortune; 5,000,000 pesetas, which today would be valued at least £25,000,000. They brought this fortune to Almería and proceeded to invest it in numerous projects. As a result, the lives of the citizens of Almería improved considerably and the Echeverría family were looked upon as great benefactors. They created the water company and installed water supplies to private houses. They invested in the first electricity undertaking and started the telephone company in the city, which ultimately extended to the whole of the Almería province. They also founded a construction company which was responsible for the building of the Dique de Levante – that is, the breakwater and port of Almería.

The family lived well and bought several real estates, such as La Pipa and Hoya de Molinero; a huge *hacienda*, high up in the sierras. Here, their daughters would continue with their piano lessons. In the summer, the grand piano was transported on the backs of mules.

They would go to Madrid for the opera season and were guaranteed a theatre box at the Royal. Their guests were entertained lavishly... nothing less than silver service! Wedding and baptism celebrations were entrusted to Lhardy of Madrid. They would deal with the whole thing and despatch everything by train to Almería, but the train would have to stop especially at a halt specifically constructed for the use of La Pipa.

The family had a kindly attitude towards their servants and were constantly preoccupied with their welfare; consequently they were much respected, even loved. Guillermo's wife had a strong resolute character which she instilled into her children, as well as making them God fearing and ensuring each cultivated a social conscience.

Those were the years in which the Hijas de Caridad (Daughters of Kindness) had undertaken the task of setting up the *manicomio* (asylum). The Mother Superior, also from Navarra, was called Sor Policarpa and she and Ana had been childhood friends. The Mother Superior was a powerful woman and encouraged Ana to support her in her social work. This religious order not only took care of the asylum, but also worked very hard to improve the lives of the inhabitants of Los Molinos.

Pedro López Rull was born in 1853. As a cadet at the Academy of Infantry, he was despatched to Aragón and Navarra. He assisted in the defence of *Villa de Guetería* as a lieutenant, rising to the rank of captain as a result of his meritorious war efforts. In 1877, he went to Almería in order to marry Trina Bocanegra; a descendent of Admiral Bocanegra, who assigned his fleet of warships on the side of King Alfonso XII for the conquest of Almería.

Nicolas López Rull was the medical officer for the local government and for the railway company, Ferrocarril Linares-Almería, built by his brother Guillermo. Nicolas married Carolina Viva and they had one daughter, Concha, who died in 1931. The whole family was devastated by her sudden death from a mysterious illness.

Both brothers distinguished themselves by their valiant military careers as Isabelino supporters. However, the family suffered considerably in the war in Cuba, as did all of the nation at the loss of the colonies.

The war in Cuba began when the USA blamed Spain for the loss of one of their battleships, which blew up whilst anchored in the harbour in Havana, the capital of Cuba. The account of this incident was no less than a deliberate excuse by America to flex its muscle power at a time when it was looking for expansion, and had judged that Spain would be at its weakest point of her imperial power, following many years of strife brought about by the long and bitter Carlist Wars. It was a shameful act by the USA, which robbed Spain of many of its colonies, such as Cuba, Puerto Rico and the Philippines, to name but a few.

The brothers were always united by their affection for each other and their families generally got together for important functions such as a hunt, the *corrida* (bull fight and *fiesta*) and social gatherings, which were an excuse to lay on a big spread. These over-the-top functions proved to be the family's weakness and to some extent, their downfall. No expense was spared; their coaches frequently made journeys to Madrid. They would often take their own coach to Huercal Overa, and then continue their journey by train to Madrid.

They worked and dreamt Almería. They were great politicians as well as military men, architects, doctors, notaries, lawyers and senators of the state. It is obligatory for visitors to Almería to view Los Depósitos de Agua and, of course, La Pipa because we have been told by our ancestors, time and time again, of the pleasure they received from playing together in those wonderful gardens.

Chapter One

As a small boy, it never occurred to me that we might be living in utter luxury or that we were enjoying such a privileged position, for I had known nothing else. Although at times I did wonder why some people, particularly our servants, always called me 'Don Emilio' (as in 'the honourable Emilio'. Incidentally, the full name given to me at baptism was quite a mouthful... Don Emilio Julian Leandro Guillermo Mateas López López Echeverría de Peralta).

On reflection, it should have been obvious to me that there was also great poverty and country folk had a very hard existence, but at that time, it was status quo; a job with our family was greatly prized. It meant security for life and probably not only for the employee, but probably for the rest of their family too. Caretakers were employed at each of the properties and the butler, footmen, cooks and some parlour maids and personal maids moved in with the family, according to the seasons. For example, a retinue of staff would accompany our family to Madrid for the opera season.

Grandfather led a busy life, having such a variety of businesses to run. Some of his companies he inherited from his father, but the majority of them he created and developed himself, along with his brother, Enrique; the latter being responsible for many of the magnificent buildings in and around Almería. My grandfather not only introduced the telephone to Almería, which was subsequently sold to the Bell Telephone Company, but he also built the port of Almería. Grandfather inherited a bank in Pamplona, which was later absorbed into the Banco de Bilbao. He and the family at large had many businesses dotted around the north of Spain, particularly in Bilbao, Pamplona, San Sebastián and Peralta.

To my knowledge, all of the businesses prospered and were given a large injection of capital when my grandmother inherited a huge fortune on the death of her father, my great-grandfather, the Marquis de Patrullo, who died in 1888. Grandmother received 5,000,000 pesetas, which in pounds sterling at today's value, would be somewhere in the region of £25,000,000.

★

Some years ago, my wife Helena and I, together with our three daughters, had a holiday in Costa Cabaña, a seaside resort just a few kilometres from Almería, and we met a gentleman called Señor Gerónomo at the local sports club. We discovered that he had known my family intimately in its heyday and in fact told us that he had once courted my aunt Luisa. He reminisced that at the time she had been living with my grandparents, and that she had been very beautiful.

However, unfortunately for Señor Gerónomo, his intentions towards my aunt Luisa were not favoured. The family did not consider him suitable for Luisa, even though he was a wealthy man owning his own large estate near Almería. Sadly, Aunt Luisa did not manage to resist the pressure from the family and did not pursue her relationship with Señor Gerónomo. A few years ago, before her death, I met with Aunt Luisa and she spoke to me about Gerónomo with such tenderness that I was in no doubt that she had a held a torch for him, after they parted, for the rest of her life. Luisa never married, despite having many suitors. Even at 80 years of age, I could see that she had been a very beautiful woman.

★

The extent of the fallout from the first bombshell that we as children experienced was enormous, particularly for me at the tender age of two and a half. We had no idea or warning of the turn of events that were to follow, particularly as we had had such a sheltered existence. We had very little contact with the rest of the family. Nanny was the only go-between. We met the parents on specific occasions and for these, we were generally spruced up. Only at bedtime did we make direct contact with them, when we received hugs and functional kisses on the head.

I can't speak for my brothers and sister, but I can say that the shock to my system was traumatic, so much so that the events became imprinted in my mind with a clarity not normally found in someone so young. What on earth are we doing here? I said to myself.

There we were sitting on a long bench; my sister Ana holding my baby brother Manolo in her arms, and my older brother Luis sitting between us. It was a large room sparsely furnished. Looking up I recognised the nun coming towards us and saw it was the Mother Superior from the asylum. It turned out that we were in the orphanage next door.

Father had died from tuberculosis in February 1928 and although I was only two years of age at the time, I can remember being hugged by him just before he died. Within a couple of hours or so, several uncles and aunts had arrived. In front of us, they discussed our future at length and finally they agreed between themselves to split us up. Ana went with Tía (Auntie) Luisa, who was at the time living at Dr Eduardo's house, which was near to the cathedral in Almería. Alfonso, the schoolteacher, took Luis and Tío (Uncle) Manuel, who was married to Tía Maruja, took me.

I did not see my brothers and sister again for many months and by then Manolo was back with our mother.

★

When we arrived at Villa Anita, I felt that my uncle was genuinely kind and sympathetic, but Tía Maruja was haughty and had a cruel streak. She delighted in telling me that our mother had abandoned us and worst of all, she said that our mother had gone off with a common old sailor. I remember that I was numbed with fear and sat on a potty in the middle of the bathroom floor, staring at the window. I tried to understand the situation, but could not really believe what Tía Maruja had told me. If it turned out to be true, then I was truly in a terrible plight.

My small brain was working ten to the dozen. Yes, I had for some time known that my mother was *persona non grata*, at least as far as the rest of the family was concerned. As the years passed, I learnt the reason why, but at that point, all I wanted was to go home to my mum, no matter how bad she might be. My father had died from tuberculosis at only 46 years of age and the family had blamed my mother for neglecting him during his illness. He apparently still loved our mother and could not see that she no longer cared for him. The family had pressured him to leave her, but divorce in those days was unthinkable.

Father had been very good looking, dashing and adventurous. He was thoroughly spoilt, had money and lived in a privileged position. He was adored by my grandfather, and consequently had *carte blanche* to do what he liked when he liked. He was footloose and fancy-free. He had the choice of many available young things, from Bilbao to Madrid, let alone Almería. But no! He had to fall in love with Amelia, a pretty telephone operator who worked for him and whom he refused to sack. This girl was not only socially unacceptable, but she also emanated from Cabo de Gata... At that time, there could be nothing worse!

I am glad that I did not know all the circumstances about my parents at the time we were unceremoniously dumped at the orphanage. It was bad enough to find myself stripped of all the protection I had come to know, and having to suddenly face a very hostile environment. It was terrible. I needed an anchor and had to have my happy home memories, as I had nothing else.

By abandoning her children at the local orphanage, my mother unwittingly played into the hands of the disaffected family, and on 3rd July 1929, without a word of warning, she was made to sign away my father's estate. This was probably illegal. The burning hate that the family had for my mother turned them into a greedy and malicious bunch, as can be seen by the fact that they made my mother sign the document of disinheritance on the very same day that my grandfather (her father) died. This deed of assignment she signed also on behalf of her children. Our grandfather died without knowing that his eldest son's children would be cut out from the specific bequests in his will. Much of the property and shares in the many businesses were divided up between my father's brothers and sisters. This included shares in the Bank of Bilbao. The action taken by the family and my mother effectively cost my sister, my two brothers and me a huge fortune, as well as the ancestral homes – especially La Pipa.

Gradually, Tía Maruja's hostility towards me subsided to the point that life began to be reasonably pleasant. I believe that her attitude to me changed when she learnt that she could not have children of her own. Her husband was able to point out how lucky they had been to have me, in such unfortunate circumstances! Tío Manuel had not only received his share of my grandfather's estate, but also the spoils from my father's estate. There were very visible signs of my uncle's new-found wealth, by way of the beautiful antique furniture which arrived at the house, plus chandeliers, a load of oil paintings of the family and many expensive *objets d'art* and memorabilia, all of which I was able to recognise instantly. I think that when Manuel and Maruja realised that I knew where it had all come from, my aunt started to say that bringing me up as a son was not going to be cheap. She would tell me that it was a small price to pay and that I should consider myself lucky that I had not been left at the orphanage and that I was fortunate that I had not been chosen by any of my other uncles and aunts.

Instead of being considered a liability, I was now paraded about and actively promoted. I was circulated amongst both the family at large and the elite – a little Lord Fauntleroy in fact! I was always dressed in velvet suits or so it felt, and if not, I was in a sailor's outfit. To go with my new-found popularity and as an important appendage to my uncle and aunt, I was given a pretty young nanny all to myself.

María was a godsent friend. Up until now I was expected to play on my own, only to be seen when required and certainly not allowed to speak unless spoken to. Contact with the servants was totally barred; in fact, I was not even permitted to enter the kitchen. However, in spite of my much improved circumstances, I still missed home and my pleas to see my mum were not only ignored, but discouraged.

Fortunately, I had another very special friend at Villa Anita. He was a spaniel that I called Perro. Not a very original name I know, but I loved him more than anyone in the world. He would love to lick me whenever he got the chance and Tía Maruja could always tell that I had been with him and she wasn't amused. He was not very friendly with other people, but with me he would wag his tail in such a way that I always knew what he wanted or what he wanted to say!

In the summer, poor old Perro suffered with the heat as he had a very thick, curly coat. One day, I decided that he was too uncomfortable and that I had to try and cut off some of his fur. Unfortunately, he became frightened, ran away from me and fell down one of the manhole covers that had been left open by the gardener. The manhole led to an underground irrigation system of channels leading to the home farm. Perro got stuck. I climbed down to get him, but the more I tried to pull him back, the more he went forward.

Water was regularly released from the huge *balsa* (reservoir) to irrigate the fields most evenings, after a scorching day. Villa Anita, together with the home farm, consisted of about 800 acres. It was now getting quite late so it would not be long before the water would be turned on. I was terrified that Perro would be drowned; I had no alternative but to tell Tía Maruja that the dog was stuck in the underground channel. She became enraged and told me in no uncertain terms what was going to happen to me when my uncle, Tío Manuel, returned from the office.

The farmer was summoned to rescue Perro. We could tell roughly where he was by the faint whimpering sounds that we could hear. Juan, the farmer, lifted two or three stone slabs and was able to get Perro out unharmed.

Perro licked my face all over, as I cried with relief. Perro wagged his tail incessantly. He looked more like a poodle as I had managed to cut most of the fur on his rump before he had run off! I am sure that Perro had by now realised that I had only tried to help him and was very forgiving. Within minutes of Perro's rescue, water came gushing through the channel. No one had told the farm labourer about Perro, so he opened the sluice at the *balsa* at his usual time. Tía became incandescent at the sight of Perro's fur; a severe punishment was now almost inevitable. When Tío Manuel came home, I was duly reported, but he could see only too well by Perro's appearance what I had been up to.

Fortunately for me, Tío Manuel was a kind man. He took me to my bedroom, shut the door and told me to yell my head off every time he hit the bed with the belt Tía Maruja had given him on his arrival home. I did as he asked.

★

The family coaches were kept at Villa Anita. There were four altogether; two were for long distance and bad weather. These were closed in and were in a green and black livery. They had a postilion seat high up in front and at the rear, there was a platform which took a couple of standing outriders. Whenever Manuel and Maruja went to Madrid or Bilbao, two footmen would go with them. I vaguely remember that they carried firearms, as it was not safe, even on main roads. These two coaches had the family crest on each door panel, and gold and green tassels hung from each corner of the postilion seat in the front. The other two coaches were open style, similar to a landau, black all over and were in use regularly until Tío Manuel bought himself a motor car; a Fiat, in pale blue.

Tío Manuel engaged Miguel to drive it. Going by coach was far more fun though. I preferred the horses, which were kept in the stables at the far end of the drive, leading to the home farm. Although I liked horses, I was never that keen on actually riding them, so I certainly did not take after my father in that respect. A gentle canter was the best that I could manage, even when we took them out exercising.

Life continued to be more than tolerable, particularly with the arrival of María. She was quite stunning and got many admiring glances and piropos (compliments) when she took me out. I kept quiet when her boyfriend used to climb over the wall to come and see her – this boyfriend seemed to appear from thin air! María was like a honeypot, and caused quite a number of skirmishes among her would-be suitors.

María liked me, because I never told on her when we deviated from our official route on our evening walks. She would also overlook my transgressions, particularly when I went to see my very best friend, Pescadero. He was older than me, and although he dressed in rags, I admired him hugely. He was a philosopher, a thief, a fisherman and altogether a most exciting person to be around. Needless to say, he was officially 'banned' from my life. Contact was only possible by the clandestine meetings arranged through Carlos, the son of the village grocer whom I met naturally and frequently. I did not know at the time, but I was to meet them both again in later life under very different circumstances.

I had a positive and enquiring mind, but also a dislike for some of the imposed restrictions on me, (which at times were harshly meted out) by my lukewarm substitute parents. As a result, I became a regular desperado with a zest for adventure and a spirit of defiance. This new outlook accelerated my introduction to junior school!

I was marched off to the convent at Los Molinos, which was only a mile or so away when I was four. The convent had been established at the same time as the *manicomio*. The family had been associated with the foundation of both places, particularly my grandmother, by way of money and personal involvement.

The church was a composite part of the convent and it was very beautiful, as it was the custom of the times to adorn the statue of the Virgin Mary with jewels (given by my grandmother). The jewels remain there, despite the Civil War.

★

To see men with balls and chains attached to their feet was an initial shock on my first day at school. Our playground stood between the school and the asylum. The inmates would stand close to the fence at the end of our playground. However, we became accustomed to their plight and could not help but feel sorry for them. We got to know some of them very well and even knew their real Christian names, and these men would often tell us about their families. The more violent inmates were kept in an inner courtyard, but we could hear them making weird sounds. The ones we got to know, and there were many of them, loved to take sweets from us and in return, make us things. Some of the inmates were extremely clever.

I was the youngest in my class at school, but in spite of that, I was able to keep up with them. In fact, on one occasion I was held up to the class as an excellent pupil – quite a rare event! On the whole, I believe that I was reasonably well behaved. I particularly remember Tía Maruja's surprise when Mother Superior told her that she was pleased to have me in the class and that I wasn't a problem at all. No, not at all!

My family was the only family in our part of the world to own a car; everybody went by bus if it did not break down. Or they went by pony and trap, by mule, donkey or the more affluent, by bicycle. However, sometimes, María was late to pick me up from school (occasionally waylaid by her boyfriend), so when that happened, I had the opportunity to walk home with some of the lads. The journey home might have had a few diversions but at no time was it dangerous. En route, or with a small diversion, we would pass a very large *balsa*, which we climbed. As we reached the top, the noise would be deafening; frogs would be sunning themselves at the edge of the *balsa*, croaking for all they were worth, but as we approached they would jump into the water on mass. Over the water there were usually dozens of dragonflies skimming the surface. Nearby, we were usually able to pick very juicy black *brevas* (figs) and fortunately the trees were not in view of the *cortijo* (farm cottage).

My extended walks home usually went unnoticed, as Tía Maruja would be having her traditional siesta. None would have the temerity to disturb her, no matter what!

At the weekends, tea was a compulsory event and always at four o'clock, prompt. For this very English function, María had to ensure that I washed and generally spruced up. This was the usual time for visitors to call. We often had visits from the Romeros and especially from the Ullmanns. He was the German consul in Almería and lived near Adra, on the coast. He had a son, called Enrique, but I believe that his real name was Heinrich. I preferred to visit the Ullmanns at their home, as Enrique had a wonderful collection of superb German mechanical toys.

For the next couple of years, nothing out of the ordinary took place. Builders moved into Villa López, which was high up overlooking Villa Anita. The drive, about a third of a mile, was only a few yards along the road from the huge wrought-iron gates that had the name 'Villa Anita' cast into them.

The house was renovated from top to bottom and extensively enlarged. The staircase was replaced with marble and the banisters were hand-carved in black oak. The dining room walls were made from stone blocks and were panelled in black oak, up to five feet or so from the floor and the room had a beautiful chandelier in the shape of dragon's heads. The study was very similar to the dining room. The furniture is said to have come from a galleon. A glass-fronted bookcase, the whole length of one wall of the study, contained the original Arabic works of algebra, handwritten on vellum. I believe that the Moors were the inventors of algebra and Tío Manual said to me that the manuscripts in the study contained the first workings of their new mathematical concept.

Such drastic alterations meant that the builders were at Villa López for many months. Personally, I was looking forward to the day we moved, as Villa Anita, whilst pretty large and attractive, was nevertheless a true villa; a single-floor ranch-style house, and I was now looking forward to going to bed upstairs. But suddenly, in the middle of it all, Tío Manuel was taken very ill and had to be rushed to Madrid, where his brother, Eduardo, operated on him for gallstones. On his return to Villa López, Tío Manuel looked very weak, and used to rest on a camp bed in the middle of the hall whilst the workmen were still knocking the place apart. One day, a lizard was disturbed by the workmen and ran over his face as he lay asleep.

After a few weeks, things started to go wrong and the doctor was called in urgently. The weather was very hot and Tío Manuel had been lying down without a sheet over him. He complained of a pain underneath the long scar, left as a result of his operation. I could see that it was inflamed, but then suddenly noticed what looked like a piece of cotton sticking out of him! The rest of the wound had healed nicely, but where the piece of cotton was poking out, there was what looked like pus – very nasty. Uncle was whipped off again, this time to Almería General Hospital.

Well, it was unbelievable. They opened him up again and found a swab had been left inside him, when they had removed the stones from his gallbladder. Tío Manuel had had a narrow escape, but my stock also went up visibly after that as, after all, I was the one who had spotted the likely cause of the problem.

★

In the spring of 1932, we finally moved to Villa López and my lovely Pura arrived. She had been working at La Hoya de Molinero, our old family home in the sierra, near Olula. She was a brilliant cook and kindness itself and she worked so hard, that even Tía Maruja would have to tell her to go and have a siesta.

In Villa López, my new bedroom was on the first floor with French windows leading onto a patio, which had blue railings on three sides. It had a blue and white hand-made tiled floor and a raised surround for sitting. During the day I would sit under the awning and play for hours on my own. It was a great vantage point, as on one side I had a panoramic view of the sierras, the back garden and the *balsa*. Facing the French windows, you could look down to the city of Almería, the sea and La Torre. On the third side, I could look down onto a patio, on which there was an enormous oval table made of white marble. It had wrought-iron legs partly recessed into the ground. It was capable of sitting twelve or more for a several course dinner on a summer night. Beyond the patio, there was a formal front garden and then a drive to the lodge at the entrance to the property. Villa Anita's entrance was only a few yards away, to the right. From my upstairs patio, I was able to see the rooftops of El Alquian and then the sea, to the horizon.

During these times, my brother, Luis, proved most difficult. He was unhappy and resented the way that things had turned out for us all. We still owned El Depósito, to which he was virtually banished by the family – he was supposed to have been looked after by the resident caretaker of the main house. But my brother was allowed to be loose and wild, as the family appeared to have very little interest in him. Consequently, I saw him very infrequently.

My sister on the other hand, settled down with Uncle Eduardo Pérez and my Aunt Luisa, who never married. I saw her regularly, for Eduardo was a doctor and was therefore often in demand by the family members. On one occasion, I had climbed onto some stepladders in order to lean over the raised fountain in the sunken garden at Villa Anita. I was trying to catch a big goldfish, which had caught my eye some time ago. Unfortunately, in my excitement, my foot slipped down a rung and I bit my tongue. My teeth had cut my tongue in half, leaving no more than an eighth of an inch either side. Blood spurted out profusely and I very nearly filled a small basin with it. I was perhaps, for the first time in my life, really frightened. Eduardo was able to insert eight stitches into the tongue and was of the opinion that I would have a speech impediment – this has not proved to be so!

Unfortunately, Perro died a couple of months before we moved. As you can imagine, I was very upset and no one could console me, especially not with the promise of another dog when we actually moved. However, the saying that time is a great healer proved to be right. I was desperate for Perro's replacement no sooner had we moved. Uncle Manuel insisted that this time it had to be a guard dog and not just a pet. The Alsatian bitch that we bought turned out to be very ferocious. I was the only person she would allow near her. I called her Perra, in memory of Perro!

Perra appeared to like her name and we became the best of friends. Her life was under constant threat as she wasn't averse to having a nip at anyone she did not like. Regrettably, there were many! In the end, Perra had to be chained to the tree next to her kennel. We both enjoyed a siesta under the branches of her tree. It was a great pity that she had to be chained, as it made her very angry and suspicious of everyone except me. I was the only person that she would allow to give her food and in fact, she was really only doing the job of defending her own territory. You could not have anyone more loveable.

Chapter Two

I was naturally brought up as a Catholic and was confirmed when I was seven years of age. This took place just before leaving prep school, the one next to the asylum. I can remember the morning of my confirmation very well, as I was dressed all in white; my first ever long trousers, an embroidered shirt and a bow tie. I had a sash that ran from my shoulder down to my waist. Embroidered in the middle was a gold chalice with a round wafer halfway out of it, and gold rays of light. I looked at myself in the mirror several times, as I could not get used to it. Could this be me? I looked so angelic; that could not be possible! Nevertheless it was to be a very important day. It was my first Communion. I was not allowed to eat a thing for breakfast – only water!

Religion had not entered my head, but for the odd occasion. I felt that it should be left to the grown-ups. They would constantly cross themselves and make regular visits to the church, where they would queue for the confessional – presumably they had the odd sin, but I thought they had decided that as I had so much to confess, that I would not know where to start! Religion to me of course, meant Catholicism only. I had never really known anything else in my life.

Today was important in more ways than one. I had been told repeatedly that it was going to be my day. Lots of goodies; all the things I liked best, including my very favourite *turrón*! This was to be awaiting me on my return from church. I had never realised how sickly sweet this became, when you had more than enough to get through!

I got dressed very early in the morning, so had to hang about for ages. Eventually Tía Maruja and I were driven down to the church. On arrival, I had to join the other First Communicants. We were all pretty self-conscious and tried to pretend that it was an everyday event. Mind you, I did think that some of the girls really did look gorgeous, which was so different to normal! Up to now, girls had been a closed subject to me, other than perhaps Marisol. I did look forward to seeing her always, as she was such a tomboy.

At last, it was all over and we each had our photographs taken at the entrance to the church. I remember the photograph still in a silver frame. It was to have pride of place in the sitting room over the fireplace. I think it was there for a couple of years and then it got mislaid. When I got home from church, I found a load of presents awaiting me. Mostly sweets, but also a very nice white rosary, several cards, a couple of small prayer books and the inevitable Missal, which a Catholic finds essential reading. In one of the small prayer books, I came across a lovely little short prayer, which went like this:

> Jesús, José y María,
> Os doy el corazón
> Y el alma mía.

I said that prayer by the side of my bed for many years; I suppose I did this until I considered it not suitable for a grown-up. I am sure that a translation is not necessary. The new-found fervour was unfortunately rather short-lived. The more I though about it, the more I didn't like it. It seemed to me that instilling such fear into people about being condemned to hell and eternal fire was not what God really had in mind, and it certainly wasn't the prime object when he created us. In the Catholic Church, it was assumed that hell, purgatory and eternal damnation was the order of the day... where was this love, the forgiving friend of mankind, and the children?

Things reached breaking point when my saintly and beautiful cousin, Conchita, died suddenly from some virus which, at that time, the medical profession knew very little about. She and I were of the same age and had been very close friends. She had long fair hair and bright blue eyes, a throwback to Sarah Anne Mills, my great-great-grandmother.

The whole of Almería was devastated by Conchita's death and the family, utterly disconsolate. Everyone went into mourning. The funeral was huge. Her little coffin was borne on a white hearse pulled by four grey horses. Each horse had white plumes on their heads, similar to the Prince of Wales feathers. At the cemetery, on the outskirts of Almería on the road to Granada, I was left with the other children, a short distance from the graveside. We sat on a low wall surrounding a pit of about twenty by fifteen feet. Moving about inside the pit was a tall, thin man dressed in tight-fitting black clothes. He had a long tail and wore a mask depicting a ghastly and terrifying face, of the Devil. We were all visibly distressed and many of the younger children cried. To this day, I do not understand the reason for this ritual and the necessity of it being so frightening to us. It was cruel to say the least!

★

When we moved to Villa López, the builders went down to Villa Anita and carried out extensive renovations there. Within a few weeks, it was let to Major Websdale, who was a colleague of Tío Manuel and a co-director at the electricity company, Fuerzas Motrices del Valle de Legrín. This was now largely owned by Whitehall Securities, who had their office opposite the Cenotaph in Whitehall.

Tío Manuel was a chartered construction engineer, with a degree from Madrid University. Major Websdale, better known as 'Webby', was a chartered electrical engineer with a degree from London University. Between them, they built the hydroelectric scheme up in the sierras, which supplied a large chunk of the province of Almería, as well as Almería city itself, with electricity.

I took to Webby immediately and considered him *muy simpático*. He spoke perfect Spanish, so most people who met him for the first time really thought that he was Spanish! I would walk down to see him quite a lot – in fact, most days, especially at weekends. Tía Maruja thought perhaps that I was becoming a bit of a nuisance to him and tried very hard to stop me. Webby missed me if I didn't go down to see him and, at first, became quite upset when I didn't go, until I told him why. Webby was the man who eventually was to become my guardian, but more about that later.

Very shortly after Webby's arrival, a friend of his, Alonso, came to live with him at Villa Anita. I didn't know what he did, but he appeared to be well off. Alonso's father had at one time been a Spanish consul in one of the South American countries. His mother, a very petite woman, was mother to 21 children, all of them having survived into adulthood. She was very vivacious and full of go. They all got on well with Tío Manuel and Tía Maruja.

Our friends at Villa Anita proved to be greater providers of toys and books than my own family were, but their moments of generosity were few and far between. One day Webby, on returning from one of his regular trips to England, gave me a 40" yacht, with adjustable sails and an auxiliary motor. Never in my life had I had such a splendid present. If I was not sailing it in the *balsa*, it would be kept on top of my toy cupboard; this was the pride of place for such a magnificent object.

All the servants lived in, accept Angeles who came most days and lived at Cuatro Caminos. She was a very hard worker and needed no supervision from my aunt or from Pura. She appeared to be old as she was well tanned and had a multitude of wrinkles on her face. She was always well dressed in black and I think that made her appear older than she probably was. She washed our clothes in the sink outside in the courtyard, which you could approach from the kitchen and the back garden. Angeles had no hot water, so the clothes were rubbed and slapped on a corrugated concrete draining section at the side of the sink. Even the hot water in the house during the summer was by courtesy of a large water tank on the roof, which the sun heated to a temperature good enough for a hot bath. Temperatures of 90 to 113 degrees between June and the end of September were the norm.

Although I was not allowed to talk to the servants, I could nearly always have a chat with Angeles as she invariably worked on her own in the house or outside. I found her most interesting. Even though she could not read or write, she could tell me a lot about the countryside, about the people, their superstitions, and of course the wildlife, which was far greater than I had imagined. I knew all about the wolves up in the sierras, as they would, on occasions, come down to the village at night. They came into our back garden sometimes and caused havoc amongst the chickens, pigs and goats.

Angeles lived in a cottage between the village and our grounds. Perhaps the description 'a cottage' is not altogether accurate though, for the front was like any conventional whitewashed cottage, but when you entered, you were in a cave hewn out of rock. It was quite large and wonderfully cool. There was an air shaft at the back, where she kept big casks of home-made wine and several cloches in which she farmed giant snails – the variety we call *escargots*, if you want to be very posh! When I was little I loved them, but I can't face them now. On more than one occasion, I went home full of snails and home-made bread and feeling somewhat worse for wear from the effects of her liberal tots of strong wine. María, my nanny, would keep me out of sight until I had slept it off. Mussels have now replaced snails in my life... Nothing like *moules marinière* – my wife Helena's favourite also.

Because Angeles had the keys to all the rooms that she was responsible for, I had to take the opportunity of having a good scout around when she did the cleaning. The rooms were normally kept locked, as there were many valuable objects and silver ornaments lying about. One day, she let me go into a storeroom in which old furniture, pictures and discarded equipment, carpets, hats and old-fashioned clothes were being stored. I came across my great-grandfather's ceremonial uniform, which he wore as commander-in-chief of the army in Cuba. It was a very splendid uniform and I recognised it as the one my great-grandfather was wearing in the oil painting of him, hanging in the dining room. It was hanging next to the painting of my great-grandmother. Both paintings had been done by Molina, a well-known Spanish portrait artist.

It was fortunate that, during one of our visits to Villa López with my wife and children, I decided to take a colour photograph of both portraits, as my cousin Mercedes has now taken them to Vienna, where she lives with her husband, an Austrian count. The family was very upset that Tía Maruja had left the entire estate to Mercedes, her blood niece. All of my father's assets and heirlooms, the López Echeverría inheritance of which Tío Manuel was the trustee, should have been passed on to me, my sister and our two brothers. Instead, he broke the trust when he died and left everything to Tía Maruja, his wife.

When Tía Maruja was about to die, she called Mercedes to her bedside and told her that she wished to see the family solicitor quickly, as she had realised she had made a terrible mistake and wished to change her will. Mercedes assumed that she had everything to lose, so did not come back until Tía Maruja had died. My brothers and sister are now of the opinion that Mercedes had been Tío Manuel's mistress for quite a few years, and that also, he had written me off when I did not go and see him just a few days before he died from cancer.

I was blissfully unaware as a small boy that I was to lose my inheritance. But, subconsciously, maybe that was why I felt more at home with my constant companion, Perra, than with the rest of the family. Perra and I were very close and a strong bond developed between us. Communications with my uncle and aunt were pretty thin at the best of times; only at meal times did we have any sort of rapport and, even then, it would be me answering their questions. I was not allowed to bring up a subject, and certainly not allowed to ask questions. Fortunately, I did have María, Pura and Angeles to talk to and, to a lesser extent, Juan. Otherwise I would have had no idea as to what was going on. I was about to be taken out somewhere, or someone was coming in to see us and I was always granted the minimum of information, which invariably ended up with a stern warning as to whom might be my only source of help if my behaviour left anything to be desired.

'Now you will behave yourself, won't you?'

I quite liked socialising if the right people were there. Some were terrible bores, but Marisol made all the difference. She was the middle daughter of a very wealthy banker and wine exporter called Señor José Romero Balmas. He and his family lived in our old house, La Torre. I am not sure whether they bought it or just rented it from us, which in any case would have excluded the home farm. Señor Romero was a Freemason and this was to cause him a lot of problems later on.

Generally we went to church on Sunday mornings to Plaza Vieja and strange as it might seem, I looked forward to it! After Mass, Tía Maruja would buy a huge quantity of *churros*, which were deep fried outside the main entrance of the church during the service. *Churros* are made of batter, extruded through a machine into long sticks and then deep fried in boiling olive oil. They are then sprinkled with a load of sugar. Most people would dunk the sticks into a large mug of hot drinking chocolate. Because you were expected to take Communion during the service, you were not allowed to have had breakfast beforehand, so naturally by the end, we were all starving – especially me!

Like the church at the school and at the asylum, the statue of the Virgin Mary was adorned with precious jewels donated by our family. These jewels were on permanent loan, so to speak, as families did not reclaim them unless they fell on hard times – well, that was the idea, but I am sure that no one would exercise their prerogative for fear of the shame that it would bring to them. Our jewels are still there, as I have seen them with my own eyes. This is in spite of the pillage and destruction that went on during the Civil War, especially in Communist-held territory. Tía Maruja, needless to say, went to church every day, as if she wanted to keep any eye on our jewels. She would have had to have been very ill indeed to miss Mass. Our chauffeur Miguel would take her in and wait outside for her.

We had a beautiful chalet with accommodation for about six people, at Agua Dulce. It had three double bedrooms and a veranda all the way around with one end over the sea, supported on stilts. I believe that it belonged to the electricity company and was for the use of the directors and their families. Webby had a chalet of his own at El Alquian, which was much nearer. It was very nice, although smaller than our chalet. Webby, Tío Manuel, another English director, Will Smith, and I would go to it most afternoons. They would work from 7 a.m. to 2.30 p.m. and then we would all got to the beach and have a barbecue lunch. This avoided having to be indoors during the oppressive heat of the afternoon. Generally, we went to Webby's chalet during the weekdays and to Agua Dulce at weekends, where we were able to meet the other English families.

I found the English element interesting and frankly, exciting, not as it is generally believed. Webby was a super bloke as far as I was concerned, and then there were Mr and Mrs Smith, with their two daughters. The Smiths were a very warm-hearted family and we often paid them a visit in Almería, where they lived. The eldest daughter was the more attractive of the two, and she was full of life. She would continually shock the establishment as she more than often wore very slim trousers, which was not the done thing at the time. Edna did not allow this to upset her style as she liked being in the limelight and it suited her engaging and infectious attraction to other men. She had a very happy-go-lucky approach, which even the women came to admire. This was in sharp contrast to the sedate, serious and rather proper, aristocratic respectability of my family and their ilk.

The English contingent were nevertheless looked upon as being very avant garde; something to be feared, but secretly admired, as they were most definitely far more exciting than the locals, who were several light years behind. Our family were probably not quite so bad as they had always admired everything English, since the advent of my great-great-grandmother, Sarah Anne Mills, whose North American family was able to trace its roots back to North Devon.

Before long I was packed off to boarding school. I was now only seven and a half years of age, and had not had any sort of mental preparation as to my future life far away from home. I found myself as a boarder in a school of the highest credentials, certainly equivalent to any of the first half-dozen top English public schools. It was called *El Colegio de los Salesianos*, in Málaga, southern Spain. Tío Manuel's chauffeur took me to Málaga along the very dangerous narrow road, skirting the coast, to drop me at school. These days, the road is a motorway, but it is still dangerous.

When I arrived at school, I was taken to the headmaster's office near the entrance to the college. He was a most amiable and kindly man, dressed in his monk's robes. He made me most welcome, but without batting an eyelid he said, 'You must now meet your brother, Luis.'

I was absolutely stunned. I had no idea that Luis had preceded me to that school. Well, you can imagine I was delighted as it had been a long time since we had last met. Luis, however, appeared to be less than indifferent to my arrival, which I found hurtful. I could hardly recognise him as he had grown so tall and big.

The monks were friendly and approachable. If you have ever been to Buckfast Abbey in Devon, England and had the opportunity of meeting some of the monks there, you will appreciate how kind, understanding and warm-hearted they are. Their love for others is second nature to them, although I was to learn from my wife that when she was a little girl and went to visit her grandmother at Buckfast, she was always made to cross the road whenever she saw a monk approaching as they were looked upon with great suspicion by the locals, who were mainly Methodist.

'Don't ever talk to them, they are Romans,' my wife was told. Well, she had half expected them to have horns coming out of their heads. The dreaded Romans were busy rebuilding their abbey on the original foundations. Today, the abbey is a tourist attraction and consequently has brought considerable wealth to the local area.

My new school was founded a couple of hundred years ago. It was built in squares, one inside the other, with an inner courtyard that we used as a huge playground. This was capable of taking 500 or more boys at any one time. The school consisted of about 1500 boarders from the ages of 7 to 18. The inner square had columns on all four sides, with a covered walkway behind them. All classrooms led from this walkway on the ground floor. On the first and second floors were the bedrooms, each having 4 to 10 beds, depending on the age of the boys in the room. A few boys had individual rooms in a separate building a short distance away, which you could approach through an archway that traversed all the squares. These boys were in the upper fifth and the sixth forms and I hardly ever came across them.

It was well organised, considering the size of the place. In addition to the boarders, there were about 300 or so day boys. I have always considered the number 44 lucky – as that was the number allotted to me at registration. When boys left the school, their registration number was reallocated to a new boy.

My most vivid recollection was of an exquisitely beautiful chapel, which not only served the school, but also acted as the parish church for this particular district of Málaga. It was a large church, but not quite big enough to take us all at one session, so we had two Masses every morning, with sufficient space left for the parishioners at each service. The older boys had to go to the 7 o'clock service – I was able to stay in bed for another hour!

Music and singing were very important and although I was very young, the plain chant of the monks and the school choir when High Mass was celebrated made a deep impression on me. It was not long before I was asked to join the choir as a soprano. This was great, as I got out of quite a bit of study time when we had choir practice, which was held on Wednesday and Saturday evenings and of course, we were always given sweets after rehearsals!

I was to be at the Salesiano College until April 1936. I cannot deny that I thoroughly enjoyed my time there. After many years I still have fond memories of the place and those gentle men who were always so cheerful. Without a doubt, their sense of happiness proved infectious and an inspiration to us all. I am sure that they were responsible for the hopelessly optimistic trait in my character!

I recollect that I used to do well at exams. This did not endear me to Luis. Matters were made worse when the housemaster tried to spur him on and suggested that he should take a leaf out of my book and try harder! It must have been very galling for him; after all, he had far more brains than me. The problem was that he had been very disturbed by the family upheaval. It had affected him far worse than me, as he was just that little bit older when it happened, and consequently was able to understand fully what went on. He is also a very proud man and resents not being able to demonstrate his noble birth with the trappings that you would expect from anyone with such a background.

Luis was a very naughty boy at school. He became a boarder at the age of eight, but even at this tender age he had the strong urge to survive and generally behaved as a hyperactive attention-seeker, but nobody made any attempt to understand him. He would raid the kitchens at night and often escape from school and go swimming in the sea. Unfortunately he was caught by a prefect as he was trying to get back to the school and was scolded by the director of the college, Don José María Doblado. Don José had him locked up in solitary confinement. On another occasion he ate shoe polish in order to make himself sick so he would be sent to the infirmary where the food was considerably better.

The lack of love and the harsh treatment he received left deep wounds in Luis, but in spite of his afflictions, his love for our mother never diminished, unlike his love for me and for our sister, Ana. I tried to be kind and help him at school whenever I could, but we were in different classes and our dormitories were far apart.

Unfortunately, one morning in April 1936 – I cannot remember the actual date or even the day of the week – I was summoned by the director, who told me that I had to pack and leave the school at once. I was to go home with Webby's chauffeur, who had been sent to collect me. In many ways I was not altogether surprised at having to leave in such a hurry. Although we lived a pretty sheltered existence as boarders, a few of the day boys that were still able to get to school brought with them stories of political unrest and all sorts of civil disturbances and even of murders!

I had no time to see Luis or, more correctly, I was not allowed to even say goodbye to him. I tried to persuade the director to let Luis come with me, but it was to no avail. So with deep sorrow I had to leave him behind. It was to prove a great heartbreak, for I was not to see him again until August 1963.

Chapter Three

Luis and I failed to trace each other for many years, as the Civil War in Spain went well into 1939 and no sooner had that finished than the Second World War started. The journey home with Pepe from the school was very frightening as we kept being stopped and searched by marauding gangs at lonely sections along the main coastal road. Pepe, the chauffeur, was very good and told them that it was more than their lives were worth to harm us. We had to stop and fill up with petrol after a couple of hours on the road; there were no filling stations open, but fortunately Pepe had two large cans of petrol in the boot. As we got to Adra, more ruffians barred the way. This time Pepe had great difficulty to get permission for us to proceed. There was no alternative but to get out of the car and go to speak directly to some self-made chief who wanted to know who I was; particularly as I was being chauffeur driven. Pepe had a very vivid imagination and told me to let him do all the talking.

We were confronted by an ugly-looking brute, whom you could only describe as a brigand. Pepe told this horrible, hairy specimen that I was the son of a very important Englishman and that I was being taken to the British Consul in Almería. Well, that did the trick and we were sent on our way. We were now very near to Almería, in fact, we had just passed a small hotel-cum-restaurant called the Eritaña and I remember having been there to dinner. Their speciality was langoustines… yummy! Once again some character wanted us to stop. This time Pepe thought that I might be recognised and could therefore not use the same excuse. There was nothing for it but to make a dash for it! Phew, we made it, but not before the car suffered a few bullet holes at the back, and one through the rear window.

I do believe that Tía Maruja was pleased to see me and, certainly, Tío Manuel could not disguise his relief at seeing us back safely, especially after he saw the bullet holes. Pepe then returned to Villa Anita, where Webby was waiting. Within minutes, Webby came walking up, with a great big smile on his face. He gave me the biggest hug possible without actually killing me! Oh, I was glad to see him; I had missed him a lot and I didn't realise how much until that very moment.

I had not been home for long before I began to realise that things were going from bad to worse. Everything was in uproar. You could not rely on everyday things to be on time, or to be there at all. I was nearly 10 years of age now, and I could comprehend more than I was given credit for. As a result, I overheard many discussions which were definitely not meant for other ears, and some of those discussions were of a very sensitive nature.

It was becoming obvious to me that Tío Manuel was more involved in the Civil War and that he was a very important political figure, far more so than I had thought. He was keeping company with some very senior military people. No doubt at these clandestine meetings, plans were being formulated that, if they came to fruition, some very dramatic consequences were bound to follow. Several high-ranking military figures visited Villa López at regular intervals and always at night. I was generally able to overhear the conversations, by lying on my stomach on the patio floor outside my bedroom. This overlooked the patio below and the huge marble table, around which they had their discussions.

I could see that Tía Maruja was becoming more agitated by the day, especially after one night when a couple of young men arrived in a *camioneta* (small lorry). They were dressed as farm labourers, but when I heard them speak it became obvious that they were no such thing. The lorry was piled up with a load of alfalfa, or so you might think! After a few hushed words with Tío Manuel, they began to unload the alfalfa and just below a couple of layers, I was able to see dozens of rifles and also what looked like a machine gun.

Outside the sitting-room French windows, there was a garden table and chairs where we would often sit for afternoon tea. The table with its umbrella was directly over a large manhole cover. We have four such manholes; one was placed at each side of the house. They were the openings to large underground cellars and each was partly filled with water. Whenever the house was struck by lightning, which was quite often, the electric discharge would run down the conductor, exploding inside the relevant cellar.

The two young fellows lifted the cover from under the table and lowered the rifles, in bundles of half a dozen or so, into the cellar. Now I understood why Juan had pumped out the water a few days earlier. When I had enquired, he said that the water had become stagnant. Obviously Tío Manuel had had prior knowledge, and hence was prepared.

★

Even at such a tender age, it was quite evident to me that central government was unable to handle the situation. The whole thing spun out of control and there was a general uprising on 18th July 1936. This was the start of the Civil War. Almería was smouldering with rumours, counter-rumours and reports of atrocities being committed by all sides. The Nationalist uprising was the trigger that brought General Franco to the scene, together with the army from Morocco and the Spanish Foreign Legion.

I wasn't sure which side Tío Manuel supported. We were Royalists by tradition and sentiment, but were most certainly not Rojos, who unfortunately controlled the major part of Almería and the province. The Rojos were predominantly Communists, as well as a sizeable element of anarchists, *sindicatos* (unions) and also a motley rabble made up of a multitude of political parties and splinter groups. This rabble came under the unifying banner denouncing the Fascists, the Carlists, the Church and, last but not least, the nobility, who were also largely the landowners as well.

Law and order went out of the window. The Civil Guard, with honesty and loyalty as an important part of their code of conduct as well as having a long-standing tradition, were the most respected of police, and they were split down the middle.

Will Smith and Geoff Websdale received the following letter from the rear admiral in charge of Gibraltar:

8th August 1936

No. 580-3

Dear Sir,
I have received from the Commanding Officer of HMS Basilisk, a report of the proceedings of that ship whilst at Almería, of which is an extract:-
'The fact that the collection of embarkation of British and other nationals was carried out without undue haste and without a hitch, was very largely due to the excellent work of Messrs. WJ Smith and Geoffrey Websdale of Whitehall Securities Corporation, 47 Parliament Street, London SW1, who were of the greatest assistance to the British Vice-Consul and myself.'
I would like to express to you and to Mr Websdale my highest appreciation and gratitude for this very valuable assistance rendered to one of HM ships under most difficult and trying circumstances.

Yours faithfully,

(Signed JM Pipon)

Rear Admiral

Webby arranged for the British Consul to function within the grounds of Villa Anita and the Union Jack flew from the highest part of the roof. I saw Webby a few times, but he and Will Smith were very busy trying to arrange for the safe passage of British subjects. They also made strenuous efforts to save some prominent families of Almería. Webby tried to persuade the Bishop of Almería to leave at the same time as some British families, but without success. He made it clear that he was in no way going to desert his people at a time of crisis.

'Tengo obligación con mis diocesanos, les he dicho, justificando mi negativa a marcharme.' (I have a duty to my people in the diocese and I have told them that I am justified in refusing to leave.) 'Pueden destruir este cuerpo, pero no pueden hacerme daño.' (You can destroy this body of mine, but you can do me no harm.)

He, as it turned out, was more fortunate than most, for he survived. The bishop's remarks were noted in *A History of Almería* by JA Tapia. 'Official figures after the Civil War for executions and murders of religious people alone, were horrific: 12 bishops, 283 nuns, 4184 priests and 2365 monks. Between 18th July and 1st September 1936, there were 55,000 executions and murders. This included women and children.'

With such a horrifying number of atrocities, one can only now understand why Webby and Will Smith tried so hard to evacuate as many vulnerable people as possible onto British warships. There were many errands of mercy by the Royal Navy operating from their Western Mediterranean base at Gibraltar.

On about 20th July, I went into Almería city with Tía Maruja in order to see her mother, a very old lady by then, living on her own except for a faithful retainer who had been in her service since leaving school. Tía Maruja was very concerned for her mother's safety as she lived a few yards from the *paseo* (high street). Huge crowds had gathered down the whole length of the *paseo*. They were in a very volatile mood and shouted repeated obscenities, waving their fists at the same time. They were occasionally interjecting with pretty derogative slogans, aimed at the landowners, the middle classes, and most of all, the Church.

Tío Manuel and the mounted Civil Guard advanced slowly but positively. At first the crowd drew back and left a path for them. Up to that point, it seemed as if the crowd had been content to let my uncle and his men pass by. Unfortunately, someone grabbed the boot of one of the guards and tried to pull him off his horse. That did it. The guard brought his sabre down with a mighty swipe at the man's arm... it was horrible. Blood everywhere. The man was trampled on, first by the horses and then by the seething mob. The guards became separated; it was a case of every man for himself. I could see that Tío Manuel was not using his sabre, unlike most of the others. He was trying to push through, and at first I thought he was going to make it. But they were hopelessly outnumbered and were no match for the heated throng, estimated to have been about 3000 in number, and most of them brandishing an assortment of weapons, including knives and clubs.

The crowd was baying for blood. Nothing else would satisfy them, so the inevitable happened. A more audacious ruffian stepped forward and grabbed Tío Manuel's left leg. He was pulled down to the ground.

Tía Maruja screamed. My heart just sank. I couldn't speak.

I went back on my heels, but not for long, as the crowd behind us were pushing forward and took us with them. What followed was nothing but confusion; people shouting and rushing about trying to avoid sporadic shots coming from some of the windows and roofs of the buildings on either side of the *paseo*. My nanny, María, was very hyped up with excitement and I had the distinct feeling that her sympathies lay with the crowd. Although you could see that she was wrestling with her natural feelings for a moment or two, she suddenly got hold of me and shouted to Tía Maruja to follow her as she pushed forward. It was quite miraculous, for we slipped into a side street within a minute's walk from Tía Maruja's mother's house.

We found Tía's mother in a terrible state. Only a few minutes before we had arrived, a posse of rather dubious characters had taken away her unmarried son, who had been staying with her. Tía decided then that she would take her mother to her sister Blanca's house nearby, as it would be safer for them there and at that point she could try and find out what had happened to Tío Manuel. We went together and fortunately encountered no more problems on the way.

María and I then set off for the long walk to Villa López, but it was quite obvious that whichever way we went in order to avoid the crowds, we were going to have to go west to east across the city. We were afraid to go too far into the suburbs, as we might be set upon by the odd self-styled militia if we were found on our own, particularly with María, who was very young and attractive.

We had to push through the less concentrated crowd who appeared to be anticipating more action. The way home was ghastly; there were bodies littered about in grotesque and terrifying poses. But the worst was to come as we approached our local parish church, the asylum and my old infant school at Cuatro Caminos.

There were men and a good sprinkling of women in the act of

chasing, killing and mutilating the bodies, some of which were nuns. I did my best not to look. Not only did I not want to recognise any of them, but I was also afraid of catching someone's eye. This vision has never gone away.

We arrived home, where Pura had been anxiously waiting for us. Very shortly after, Tía Maruja arrived by car. Whoever it was in the car with her just dropped her off and sped away as fast as he could. It appeared that Tía's mother and sister had thought it would be better for her to be at home in case someone wanted to get in touch with her about Tío Manuel.

Ultimately, somehow, I managed to go to bed, but before doing so I went into the study and found Tía Maruja on her knees, praying with a rosary in her hands... tears covering her face. She looked very frightened, was white as a sheet and was shaking all over. Neither of us spoke. I put my hand on her shoulder for a moment or two, to try to calm and console her. I decided that it was best to leave her to her own thoughts.

Next day, we listened to the local radio for any news. It consisted of a constant flow of short bulletins, interjected with frequent requests to keep in touch. As time went on, it became patently obvious that the Reds had taken control of the local radio station. It was a very one-sided affair, but by broadcasting their achievements only, you could work out how much of the region was under their control.

It would appear that the nearest Nationalist forces were at Sevilla, Cádiz and Córdoba. They were a long way off, so we felt and knew that we were well and truly cut off, with not a lot of hope of being relieved. We felt doomed... We were but a few, known for their wealth, property and Royalist sympathies. We had no word about Tío Manuel and had to assume that he had perished. Yet on the other hand, he had been a popular mayor and was generally known for his good works and sympathy for the underprivileged. We hoped, therefore, that the proletariat would spare his life, if he should have survived the lynching that we had witnessed.

We were about to have a late breakfast when we noticed a crowd of militia men, mostly sporting guns and an assortment of weapons, walking up the fairly steep drive towards the house.

When they arrived, they dragged us out of the house and made us line up out in the front. Tía Maruja, María, Pura and Juan, the gardener/keeper of the lodge and I all lined up by the entrance at the bottom of the drive.

We were subjected to a torrent of abuse. One of the ruffians had been carrying a very large metal container which he rested on the ground. I could smell that it was full of petrol. He was particularly ugly, unshaven and sweaty. He went to Tía and told her that after they had looted the house, they were going to set the place alight.

I was under no illusion; this mob intended to do us harm, and they would enjoy doing so. We were all but minutes from doom, especially after what we had seen the day before on the way home. Life had become very cheap indeed and we did not think for a moment that they would spare any one of us… Mercy would have been out of keeping in their present frame of mind. The posse consisted of a motley collection of villagers, peasants and criminals that had been released when the prison just outside the Barrio Alto had been stormed. I was already dumbfounded. I could just not believe my eyes; there in the front was a person that I knew well, a villager that I thought was a kind man, pointing a gun at us.

It was at this point that, in the corner of my eye, I noticed this slightly built figure walking up the drive. Yes… it had to be.

It was Webby.

This Englishman in grey flannels and an open-necked shirt, cool as a cucumber, went up to the horrible specimen who appeared to be in charge. In a trice, he was engulfed by the whole mob. I suppose that there were about 40 or so of them, and without a doubt they were the most ugly, unpleasant and fearsome bunch you would ever wish to meet. Their eyes were full of hate.

Suddenly Webby spoke. '¿Qué te pasa, hombre? ¿No tiene razón, ni verguenza? Deja esta familia tranquila… ¡Fuera de aquí, sin parar!' (What's happening, man? Have you no sense or shame? Leave this family in peace… Be off with you!)

Like a group of small children, they slunk away without a murmur. It was as if by magic. I stared in utter amazement. I looked at this hero with incredulity. I believe that this was the moment when I knew that our destinies were to be inexplicably as one. Here was the man who had won the Military Cross, was mentioned in dispatches four times for bravery in the First World War and once in the Second World War for distinguished service.

Webby, incidentally, was also the last British officer to leave Odessa in the Ukraine, a port in the Black Sea. He had been serving with the British Expeditionary Force, which had been sent to aid the white Russians. He left on HMS *Sportive* on 7th February 1920, bringing with him the Union Jack wrapped around his body, to hide the flag from general view. This was later presented to the Boy Scouts group in Bedford. He was personally awarded with the Order of Saint Stanislaus, by General Deniken. I believe that it was a very high military honour at the time of the Czars; the nearest thing to our Victoria Cross. Webby helped General Anton Ivanovich to escape to the West. The general had distinguished himself in the Russo-Japanese War in 1904–05 and in the First World War. He organised a volunteer army of 60,000 white Russians loyal to the Czar, but was routed in 1919.

Webby was to become my official guardian, but at that precise moment in time, standing in front of our house, he was our saviour. His credentials were impeccable and I have always felt that this knight in shining armour deserved a better ward, but on the other hand, none could have loved him more than I did. I am sure that he always knew that, and he also knew that I would have given my life for him – as he very nearly did for me, my family and the rest of the household.

★

In the First World War, Webby reached the rank of lt. colonel, but came out of the service with a fixed rank of major. He joined the Second World War as a pilot officer. Being highly qualified, he became a boffin in fighter command and reached the rank of acting wing commander, but again came out with a fixed rank, this time of squadron leader. He invented FIDO, which was a fog disposal system for runways, and was awarded the OBE after the war. However, he declined to accept the honour as he considered that what he did was no more than doing his job.

His visits to the officer's mess at various airfields whilst he was engaged on guided weapons development would raise many an eyebrow, when they saw a Military Cross and other First World War ribbons, and of course, his Russian medal for valour, which had to be at the end of the row of medals, as it was a foreign decoration.

★

When the mob that had come to deal with us was out of sight, Webby said the he was mighty relieved that they had taken the hint and left as they did without questioning his authority. He had not been aware that we were in danger; he had just decided to come up to tell us that he and Will Smith would be going to Gibraltar the next day as the British Consul felt that it was now too dangerous for them to remain, as he was unable to get guarantees for their safety. It was not until Webby had been halfway along the drive that he realised that there was something wrong, otherwise he would have come up with his Union Jack, or at least armed. Personally, I believe that his sheer audacity carried the day!

Before Webby left for Gibraltar, he told me that he was sure that there would be other British warships attempting to rescue more British subjects, particularly now that things had become so dangerous. I was to keep a sharp look out for them from the roof of our house, where I could see the horizon over the sea. It was only a short gap between Cabo de Gata, at the foot of Sierra de Gata, and El Alquian.

Webby said that he would get word to the Royal Navy in Gibraltar as soon as he arrived, so that they too could look out for me. He tried very hard to get Tía Maruja to let me go with him. She felt that she did not have the authority without asking Tío Manuel first, but Webby said that as we did not have any news about my uncle that it would be better and safer for me to go to Gibraltar with him and that he would ensure that I got back as soon as it was safe to do so. Tía Maruja was beginning to waver, but I could see that she was very unhappy and afraid. It made up my mind, and very reluctantly I said that I could not leave them until we had news of my uncle. Webby was very upset, as he genuinely feared for my safety. After Webby left, I was glad that I had resisted the temptation as, for the first time in my life, I had been wanted as the only man in the house. I enjoyed the feeling for the short time it lasted... But yes, I had grown up; there was no denying that.

Webby and Will Smith stayed at the Rock Hotel in Gibraltar at company expense. Everyone was of the opinion that the Civil War would not last long, even though at the outbreak of hostilities, there had been so much bloodshed and horrendous stories of atrocities. The war in fact, continued for three years and only finished shortly before the commencement of the Second World War! Many considered that the involvement of Germany and Italy was the prelude for World War Two, in order to test equipment and to work out the strategy.

Within a few days of Webby leaving, some militiamen with armbands (I can't remember what inscription they had on them, but I think they were anarchists) came and commandeered our new car. They said that they belonged to the local headquarters and that we would get our car back when the emergency was over; no one believed them, but we were in no position to do anything about it. However, they gave us a receipt.

Miguel, our chauffeur, must have informed the authorities, as he got into the car with a couple of them without saying a word and drove it away. I never saw our chauffeur Miguel again.

We did not miss the car at first, as there was no one left that was able to drive. We felt very cut off and frankly, when the telephone was disconnected, all we could hear was sporadic gunfire, especially machine gun fire and the occasional boom of large guns.

About a couple of weeks after we lost the car, Tía Maruja had news about Tío Manuel. He was alive and being held prisoner in the bowels of a large merchant ship moored in the harbour. She was told that he was very poorly and refusing food. Tía Maruja decided then to go and stay with her mother in Almería city, so that she could try to go and see him. Her mother was only about half a mile from the port, so she could visit him regularly and bring him food. Tía Maruja was very distressed when she saw Uncle for the first time, as he had lost so much weight. Food was pretty scarce and what you could get was exorbitant... An egg, for example, cost 10 pesetas or about 50p in today's money. Tía's regular visits and a regular supply of home-cooked food made a big difference to his health. I never realised that I was never to see him again, as things turned out.

When Tía Maruja went to live with her mother, I was left alone with the servants. María was to keep in constant touch with Tía through her boyfriend, who had joined up and was stationed at the main barracks in Almería. Actual fighting appeared to have subsided in the immediate vicinity, but old scores were being settled by small groups of self-appointed firing squads. People were taken out of their homes at the crack of dawn, and driven out to some lonely spot or disused quarry, where they were summarily shot, without even having their eyes covered.

Fighting was apparently still going on in and around Málaga. We had no news about my brother Luis. This was particularly worrying, as we had heard on the radio that there had been numerous murders and executions; my old school in Málaga had been ransacked and it was reported that some, if not all, of the monks had been murdered.

No one slept comfortably at night, as we were subconsciously listening to every noise, whether from inside or outside the house. I had brought Perra into the house, not only for her own safety, but also to keep me company by sleeping at the end of my bed. She was very alert and kept getting up and twitching her ears, which did nothing to reassure me!

Although we had electricity most of the time, I did not like to put on the lights in case there was someone prowling about. Unfortunately, the batteries in my torch had packed up and I had no hope of getting more. We had several oil lamps, so each resident kept one by their side. We had a fair amount of olive oil and could afford to keep lamps burning most of the night with a shade over them. I suppose that the worst part was waiting for something to happen. Surely it would be better if we knew the worst? Sometimes I found the suspense more than I could bear, but I was the only man in the house. I had to show that I could cope with the situation, even though my stomach would be turning over. I remember that it was then that I became addicted to bicarbonate of soda.

Strangely enough, when the inevitable did happen, I was relieved even though I had no idea as to what was to follow. It was, in a way, a bit of an anticlimax.

★

It was 3 o'clock in the afternoon when four army vehicles came up the drive. They pulled up in the turning circle in front of the house, and several officers of mixed nationalities stepped out of the *camionetas*. It turned out that they were mostly Polish, and were part of the International Brigades that had been formed to assist the government. As far as I was concerned, they were Rojos!

We all came out to meet them, as I was anxious not to be searched out of the house at gunpoint. The senior officer told me that they were taking over the house and that soldiers would be camping in the garden. It appeared that we had no choice in the matter, and anyone living there at the moment had the alternative of either leaving or moving together to some remote part of the house. And that was that.

The soldiers were quite affable and apart from a few occasional demands, they left us pretty well to do our own thing. They would sometimes even ask me where they, or I, could get things for them. They thought I was streetwise… That just shows how much I had had to change in such a short space of time. The captain was really quite kind and was happy for me to borrow his field glasses first thing in the morning and late in the evening. This was in order to scan the horizon for any British warships, as I had promised Webby I would do. I got on quite well with the captain if I needed to, but generally we didn't exist as far as the soldiers were concerned.

I slept near the kitchen in one of the servant's bedrooms, opposite Pura's room. The soldiers didn't use the kitchen or the servants' area, so presumably they cooked outside. I must say that it was a relief to have them there, as I imagined that as long as they occupied our house, no one else would attempt to molest us.

Fortunately, I did not recognise any of the small numbers of Spaniards that were in this large group.

Things in Almería had settled down; in fact, it became a ghost town, as I didn't see many people about anymore. We had to be even more careful. On going out foraging for food, you would become conspicuous if you got near to any of the barricades, which were strategically placed at each end of main thoroughfares. I was now able to go and see Tía Maruja at her mother's. She was able to tell me that Tío Manuel had improved a lot, but she herself looked very sallow; the strain was beginning to tell on her. I think that the news about her brother had shocked her even further; she had heard that he had been shot for not telling the authorities the whereabouts of certain people that they wanted for questioning. Adela's husband (her brother-in-law) had also been arrested. Pérez Casinello, my uncle by marriage, was under sentence of death, awaiting his appeal. You might think perhaps that justice was now being done. The fact is that none of them had done anything wrong, other than that they were caught on the wrong side at the beginning of the hostilities. These were show trials for the benefit of the outside world, which was now taking a great interest in the Civil War. These members of my family were honourable men in public office, just doing their duty, for which they paid the ultimate price. On my first return to Almería many years later, accompanied by my own English family, I was very moved by the respect that people had for the members of my family who had been murdered – for murder was what it was. I was proud to see that many streets and public places had been named after those family members.

Chapter Four

I was always keen on the Royal Navy and Webby had taught me how to distinguish the different types of warships, especially silhouettes of the different classes of destroyers and frigates. This was particularly important as he expected any future evacuations to be carried out by such a warship, being very fast and manoeuvrable.

I was disappointed many times; they were either Spanish merchantmen or Spanish warships. Occasionally they were British ships, but they didn't stop. This went on for weeks, and I began to feel that perhaps the British had finished picking up their nationals, or that they had chosen somewhere else to evacuate them from.

Things were getting worse as far as food was concerned too; prices were unrecognisable and the black market rampant, unless you were able to show that you were one of them, which we were certainly not and, what's more, had no desire to be so. We all lost weight involuntarily and could see very little hope of improvement in our circumstances as we now realised that we had been well and truly sealed off from the rest of Spain.

When I least expected it, low and behold there was a ship, the shape of which I definitely recognised… It was a destroyer. It was a D class, but it kept moving!

At last, it stopped off El Alquian. I rushed downstairs and handed over the field glasses without saying a word, although the captain, I am sure, guessed what I had been up to. There was no time to lose! I could see that there were puffs of smoke coming from the shore, so the warship was not going hang about if it could help it.

A little while earlier, I had been over to see my mother, who was now living in the Barrio Alto suburb with some sailor or other, giving me the opportunity to ask permission to leave Spain if the opportunity arose. Tía Maruja insisted that I did that, in case my mother reported her to the local commissar, who had summary powers and therefore had no one to answer to.

I had my very small case ready packed, virtually from the moment that Webby left, so I was able to go at once after saying my fond farewells to Pura and María. I asked María to say goodbye to Tía Maruja for me. And I was off... But how to avoid the roadblocks? I had not yet worked that out. I should have done that before; I had had more than enough time to work out a plan... What a fool! There was nothing for it but to run the gauntlet. The first was at Cuatro Caminos – I ran through someone's garden and that was the first obstacle out of the way.

The next was at the approach to El Alquian itself. I had worked out that the warship had anchored opposite the beach, an area I knew well. This would certainly be a suitable place for a rescue operation as it was isolated and there were high sand dunes between the beach and the small town of El Alquian itself. I had dressed in what I thought would be the least conspicuous, but at the same time, I had to look appropriately dressed for boarding the ship. I managed to get into the town without arousing suspicion, but unfortunately, I could see a barricade just ahead of me that was blocking my most direct route. Someone shouted to me to stop, but without hesitation I ran into a side street and hid behind a garden wall. I had to lie on my stomach, as the wall was only about three feet high. I heard voices getting very close to me... My heart was thumping so hard I thought whomever it was following me would be able to hear it. One of the men suddenly said to the other, 'Deja, es un niño solo.' (It's just one boy.)

Much to my relief, they walked away.

It was quite obvious that it was my small case that had alerted the militiamen at the barricade; I was in two minds whether to dump it or not – the loss of the contents was hardly to pose a life-threatening situation, but it might look better at my hoped-for evacuation point. I held on to it and made my way down the side road, which unfortunately led to the high sand dunes.

The going was hard, as the sand was very fine, and I sank into it with each step. This slowed me down considerably, so I was becoming rather anxious about the loss of time. At last, I got to the top and to my great relief, I could see a group of people by the water's edge, about 300 yards from me. This had to be the reason for the warship to be stationary, about three miles or so from the shore. By the time I had reached the party, a pinnace (a separate lifeboat-style vessel) was approaching the beach.

There were about seven or eight men and women standing together talking in low voices, and they did not appear to notice my arrival, with the exception of one lady who turned around and smiled. She put her arm around my shoulders and gave me a squeeze. I said in Spanish (as I could only count up to five and speak about the same number of words in English) that I wanted to get on the ship as I had a very close family friend who was expecting me in Gibraltar.

A fair-headed young second lieutenant stepped ashore. He carried a clipboard with a list of names, which he proceeded to check by calling out their names. There was a certain note of urgency in his voice. Noises were coming from the direction of the sand dunes. The lieutenant was about to address my lady, who immediately grabbed my hand and whispered in Spanish to keep quiet and to leave things to her.

I did not understand exactly what was being said, but I certainly got the gist of it; yes, unfortunately my name had been left off the list, but that I was definitely her nephew. The young lieutenant hesitated for a moment, but in view of the rifle fire getting nearer, he decided to sort things out on board. By the time that we had reached our D class destroyer, HMS *Venezia*, there was quite a crowd on the beach. Those militiamen must have decided to go on looking for me. Once on board, I was taken to the bridge to meet the captain, who would decide what had to be done with me, as my compassionate lady had now divulged the truth about me… I was unable to answer the captain, not understanding what he was saying to me! The captain seemed very young, but very nice.

He said to me in Spanish, 'Tu nombre es Emilio, ¿verdad?'

'Yes,' I said, this being one of the five words of English that I knew. He told me in very good Spanish that Señor Websdale, a friend of my family and also a friend of his, had asked him to keep a lookout for me. He said that he would wireless Señor Websdale in Gibraltar immediately and arrange for him to meet me at the quay. The captain then ordered a leading seaman to take care of me. Unfortunately, I could not understand a word that he said; it turned out that he had a strong Irish brogue and that is why he sounded so different to the captain. The captain was Lieutenant Commander de Winter, who had passed out of Dartmouth Royal Navy College.

Due to the communication problems, I did not always follow instructions. The ship was travelling very fast on a zigzag course, in order to avoid the firing coming from the shore and, as a consequence, I had to climb part-way up the railings in order to avoid getting soaked or swept overboard. I was rescued by my assigned matelot, who yanked me by the scruff of the neck, and then gave me a broad grin!

Late afternoon that day we arrived at Gibraltar and there at the quayside was Webby, waving with obvious delight. I was thrilled and overcome with joy and relief. I released the pent-up tension and I had to shed unashamed tears. Here I was, in Gibraltar… my dream come true! Safe and sound. Life was now going to be just wonderful!

★

We took a taxi to the Rock Hotel, the best in Gibraltar and most luxurious, as I was to find out. It has a commanding view of the town and harbour, which at that time was a hub of activity with so much of the Mediterranean fleet in harbour at any given time. Webby had a large double bedroom, with a balcony overlooking the harbour. He had a single bed put in the room for me, as he thought that for the moment at least, it would be better not to be left on my own, in view of the recent traumas I had experienced.

I had for several years been interested in naval ships, particularly as the family had an admiral, Almirante Bocanegra. It was he who committed his ships to Alfonso XII, for the conquest of Almería in 1892. Other than his influence, Webby had also got me interested in the Royal Navy because, not only did he have a keen interest himself, but he was also very knowledgeable. It did not take long for me to learn how to read the numerous signals for the base and also from ship to ship. Some were really humorous, tongue-in-cheek sort of stuff. I was totally captivated and felt very privileged to have such an opportunity given to me. Gibraltar at this period of English history was at the pinnacle of its power and glory.

I was a Spaniard and proud of it. However, secretly, I would have liked to have been English. There was no power on this earth with such a fleet, dispersed over the four corners of the world. Not even the USA could match it. Gibraltar, at any one time, was a good place to show the sheer strength of the Royal Navy. It was not uncommon to see such battleships as HMS *Hood*, *Rodney*, *Revenge* and also aircraft carriers and countless destroyers accompanied by their auxiliary and support ships.

Some of the ships appeared to be permanently based at Gibraltar, but many were plying from one end of the Mediterranean to the other, calling at Malta, Alexandria and Cyprus, as well as many other courtesy visits to ports throughout the Mediterranean. I remember that HMS *Hood* had a very unusual silhouette in the skyline, as it had three gun turrets up front!

The Rock Hotel was an obvious rendezvous point for naval officers off duty. Lt. Commander de Winter was a very regular visitor. He would quite often give me five bob to gamble on the fruit machines for him. These machines only accepted shilling coins, a shilling being quite a lot of money for a boy in those days. I was generally lucky though, and we even shared the winnings! My time at the Rock proved to be one long holiday, particularly as Webby and Will Smith had no work to do but to wait for the Civil War in Spain to finish. The weather was wonderful and we went bathing in Playa de los Catalanes most afternoons. After a couple of weeks of this sort of life, I was relaxed and putting the recent unpleasant memories behind me.

There were quite a number of Spanish families taking refuge in Gibraltar. Some were conspicuously wealthy and several of them were staying at the Rock Hotel. There was one boy called Pepe from such a family, and he seemed to have limitless money to spend on the fruit machines... Perhaps, like me, he had not seen them before; I would watch him put two or three pounds worth of coins into the machines in one session. After some of these bouts of careless abandonment, he would walk away in utter disgust. This was my cue for a modest investment, or more likely a flutter, which came off more often than not. My view was that he had primed the rather greedy machine, and got it in the right mood for me. On one such occasion, I had the pleasure of winning the jackpot! It was five pounds worth of lovely, shining shilling coins!

You could not but realise that Gibraltar was very small indeed. I think it is about two and a half square miles in area, but what it lacked in space, it made up for in activity and exciting, interesting diversions. Naturally, I did everything that everybody did, and then a lot more! I went up to see the Barbary apes. There was a big military presence and consequently, some parts were barred to civilians. Although we could have gone into La Línea, Webby decided to play it safe, much as he liked the hotel there – the La Reina Cristina.

It was becoming pretty obvious to everyone that the Civil War in Spain was not going to end in the near future, as had been previously anticipated. In fact, if anything, it was looking like lasting for a very long time. The fighting was growing in intensity and each side was tenacious over the territory that they held. This was going to be a long struggle. So alas, Whitehall Securities instructed Webby and Will Smith to return to the UK in mid-September 1936. Fortunately, they agreed to let me come with them, at company expense.

Before we left, as a great treat, I was allowed to go on board HMS *Venezia* again and I was given a cap name band of the ship woven in gold, so that I had a memento of the ship that had rescued me from that beach at El Alquian. Luckily, the sun was just right for a photograph to be taken of HMS *Venezia*, which had the sun awnings, fore and aft. This photograph I still keep on my desk.

We left Gibraltar on the P&O steam ship *Carthage*, on 14th September 1936, mid-morning. It was a lovely day and I was able to see the Rock from the sea for the first time, bathed in sunshine, as when we had arrived in Gibraltar on HMS *Venezia*, it was very misty. I had never been on a big liner before. It was a great adventure and this time, a happy experience. I was elated with excitement at the thought of going to England, as I had heard so much about it from Webby that I couldn't wait to get there. The vivid green of the hills and coastline all the way from Mount Edgecombe, Plymouth to the Kent coast... It was so unbelievably beautiful that I honestly thought that it had been painted! It was so different to the Costa del Sol, where everything had been burnt by the sun.

We docked at Tilbury on 18th September 1936 and immediately there were problems. I had no passport, having left Almería in a hurry and, of course, unofficially. Fortunately, the immigration officer from the Home Office was clearly trying to be reasonable and allowed me to land on the strict understanding that I would produce a passport within 14 days.

The Civil War in Spain had split the country so much that even the embassies varied their allegiance from country to country. I was to prove lucky, as the Spanish embassy in London was very 'pro' the Nationalists and considered that I had been right to escape from the Rojos, so they provided me with an emergency passport, which was really just a document giving proof of identity. The Home Office was happy with it and I was accepted as a refugee on 28th September 1936.

We left Tilbury by train and after several changes, we finally got to our destination, Chingford. This was where Webby's mother, a widow, lived on her own. Her husband had died some years earlier. She was very small and slim, but rather severe looking and did not exactly welcome me with open arms... more a case of resigned acceptance.

It was bitterly cold and wet; I felt cold anyway as I was used to a much warmer climate. It was quite a contrast to the dry, sunny Mediterranean. Things were made worse by my most inappropriate clothes. The top priority was to get something warm, but right now, a mackintosh wouldn't go amiss! I had never possessed one of those in my life.

I discovered that Webby was now being called Geoff, so I decided to fall into line and probably went on calling him that for a couple of years, by which time I had a fairly good smattering of English and I decided to call him Charlie Boy instead. Eventually, this just became Charlie and unfortunately for him, he had to put up with that for the rest of his life!

The atmosphere was far from perfect at Chingford. Geoff's mother had been left reasonably well off and therefore lived a sedate life, which left her with the sole task of having to work out on whom she could inflict herself next. I am sure she kept a roster to ensure that everyone on it, got the right number of visits each year. She had many willing victims and relatives at home and abroad, the result being that she was very rarely at home.

I was definitely cramping her style by being there. Consequently, she was an unhappy girl and often complained to Geoff when he returned each day from his office at Whitehall. She became grumpier by the day and although I could not understand what was being whispered, I could tell that it concerned me. Plans were being formulated for my eventual departure, but I had not realised that it was to be imminent!

In a way, it was no surprise to find Geoff and myself at Woodford Green, on the edge of Epping Forest, as it was then. We arrived at Geoff's married sister's house at teatime. She, called Marion, was on her own when we arrived. She had obviously been expecting us, as there was a big tea prepared in the sitting room. Marion was a lovely person; I took to her right away. She made us so welcome that it just reinforced my belief that English people were by and large, rather nice. You have to remember of course, that I was rather young at the time!

The house was a large, semi-detached house and was the last one in the road before you entered Epping Forest, by a path that started where the tarmac ended. The house gave me a happy feeling and I felt in my bones that I was going to enjoy living there... I hoped that I was going to be staying. That decision, it would appear, had not yet been made, as Marion wanted to meet me first, before making up her mind. I could not really blame her for taking that precaution!

Luckily, I passed muster; she had taken to me, as I had to her. In fact, I vaguely remembered her from when she visited Geoff at Villa Anita in Spain, along with Geoff's mother and Peter.

Vincent and Sheila walked in, just in time for tea – they had been dropped off by a neighbour who had picked them up from school. The Henstridges had four children... there was Peter, about 20 years old then; Pam who was very attractive and very clever, at high school and about 17 years old. After Pam was a big gap and then Vincent, who was about eight years of age, and finally Sheila, the youngest at about six and a half years of age, and quite a tomboy.

I had to share a bedroom with Peter. He was more than happy to do so as he wanted to improve his Spanish. Vincent and I also became good friends, and eventually I went to his school. Peter was a great help as he would translate things for me and I was able to relax. And of course, it was a relief to hear my own language every evening when he came home from work.

I had no problem with Vincent over our language barrier; we always seemed to understand each other in one way or another, and a lot of signs made up for words. We had great adventures in Epping Forest, which literally started about 150 yards away from the house.

Geoff continued to live with his mother in Chingford whilst he went to work in the City. He would often come to see me at the weekends, and every now and again, he took me to London for a treat. We visited the British Museum and Hamleys. He would also take me to Bassett Lokes, the model makers. He ordered a particular railway engine to be made especially for him, Claude Hamilton, which had been in service with the Great Eastern Railway. The livery colour was a beautiful shade of pale blue.

One day we went to his office in Parliament Street, opposite the Cenotaph, and called in on Mr Walsh who was the managing director of Whitehall Securities. He was imposing, but nevertheless pleasant. He gave me a beautiful, soft red calf wallet, in which I found a new, crisp 10-shilling note. I was still a little shy, not being able to converse in English as yet, so I just smiled.

'Well, say thank you, damn it!' interjected Geoff, and that I repeated exactly, including the 'damn it!' Mr Walsh was beside himself and thought it hilarious... Obviously I had said something funny, but I knew not what!

Vincent was full of ideas and now that he had got an older friend to play with, he became more adventurous. There were times when I had to dissuade him from some of his more hair-raising schemes, which I could see might get us both in serious scrapes. We built a raft on one occasion by tying several empty oil drums together, and attached some scaffolding planks which we removed from a nearby building site. We were going to be pirates who lived on an isolated island in the Caribbean. A little imagination was an essential part of the schemes... we had no problem with this. The gravel pit, full of water, had a mound in the middle, a perfect setting for would-be pirates.

It took several evenings' work to put our raft together, as we had to hide it last thing at night. This took precious time, but we had no choice in the matter, as there was a wild gang of desperados roaming about late at night, intent on causing as much mayhem as possible. Once we were on our island, with our raft moored next to us, no one could get to us; we were well and truly cut off. The gravel pit was very deep and murky. Our island visits lasted several weeks, as we had been allowed to bring our tea with us. The first thing we did each time was to light a fire in order to brew a pot of tea. We became rather smug and complacent, and did not notice that someone had tampered with our raft. One evening, we had paddled about half way to the island, when the raft began to sink. We fell into the thick, muddy water and had to swim for it. Unfortunately, Vincent had not yet learned to swim... I dragged him along as best I could in an attempt to stay afloat. This understandable panic didn't help a bit. I was very much aware that things could go horribly wrong and I too, was getting somewhat hot and bothered.

But we made it to the bank and just lay there for a moment or two, to get our breath. Vincent coughed out quite a lot of water, but he was no worse for his experience. In fact, recounting it later, it gave him a great sense of adventure! Strangely enough, we went to the swimming baths shortly after that, so that he could learn to swim. He picked it up very quickly!

Apparently, what had happened was that somebody had taken the plug out of each oil drum and consequently, the drums had gradually filled with murky water as we paddled across to our island. We had a pretty shrewd idea who had done it. The worst thing that happened is that we got grounded for a whole week after we arrived home, soaked to the skin. Our school uniforms were a write-off. But as we were not allowed out, we had to content ourselves by playing cowboys and Indians. We tied Sheila to a stake... Well, actually the washing line post. She was not too happy about her imprisonment, but then the United States Cavalry came in the nick of time to rescue her from the Indians, and she would be rewarded with an extra sweet! She was a good sport, and didn't really mind getting the worst parts.

Pam was far too grown-up and sophisticated to have anything to do with us. I had the distinct feeling whenever we came face to face, that she could not understand why I descended to the same level as her baby brother. The look of disdain said it all, and had me frozen to the spot. Peter, on the other hand, was far more approachable and often was prepared to humour us. He had some very useful knowledge, which now and again he was prepared to impart to us. I liked him, and he was of course, indebted to me for my efforts in trying to teach him Spanish. My lack of English really reduced my lessons pretty much to practising the pronunciation of Spanish words. This gave me a big hold. I was also able to adopt an air of superiority, which I made the most of, as I had very little else with which to shine.

Marion was a very hard-working mother and in spite of having to cope with so many of us, she did her best and looked after our every need, even our individual fads. She was such a kind person that nothing was too much trouble and she always kept her cool, even under extreme provocation. She appeared to be happy with her lot, but as her husband was so quiet and reserved, it was difficult to tell. I had the feeling that she had resigned herself to a life dedicated to others and did not expect to get a reward in this world.

*

Mr Henstridge played the violin, apparently extremely well, and was a member of an orchestra in London. I never found out much about him and, as far as I knew, he worked somewhere in the City. His involvement in an orchestra was in the evening or at the weekends. He had a workshop where he repaired old violins, violas and cellos. After he repaired the violins that people brought to him, he would play them extensively, making further adjustments when necessary. I loved to hear him play; they were generally short pieces of difficult music, which would test the capabilities of the instrument he was tuning after repair. He must have been a very good musician, as even at my age, I did really appreciate the music. Most of the pieces were lovely excerpts from well-known arias and operas. Some he must have liked so much that he would play the whole piece.

In between repairs to various instruments, particularly when he had to wait for the glue to dry, he would continue making a marvellous doll's house. It was huge. In fact, I have never seen one so large in all my life! Each room was a work of art in its own right. It had electric lighting; of course, it even had a standard lamp that worked, as well as several miniature table lamps that also worked. All the furniture was accurate in every detail.

Originally the doll's house was destined for Sheila, but as time went on, it became such a grand affair and labour of love that Sheila gradually lost interest in it. It became obvious to her, and everybody else, that it was going to be an ongoing project without a foreseeable future with the children! Mr Henstridge did actually manage to finish it before he died. It was worth a fortune; it even had individual diamond pane leaded windows and many working parts. Unfortunately, I do not know of its whereabouts now; all I know is that it ended up in a museum, but I know not where.

Chapter Five

Some of my most cherished and happy memories were of holidays at Burnham-on-Crouch, where I used to stay with some relations of the Websdale family. It is on the tip of my tongue, but I just can't remember the name of the parents, even though I spent many a holiday with them. They had a son and daughter. She was the proverbial tall, blue-eyed, willowy blond who looked decorative anywhere, particularly when she wore her nautical and very casual outfits. These outfits were meant to catch her prey... and did, poor sods! They never lasted long, as she tired of them very quickly. After all, there were so many waiting and eager to take her bait! It seems quite incredible that I can only remember Paul's Christian name, but there it is. Anyway, he, the son, was an excellent sailor, keen and very knowledgeable of the estuary and the various channels... if you were under sail, absolutely essential information, as I found out to my cost on more than one occasion.

Paul's parents were keen on the sea, full stop, so much so that they sold their house in London and bought an old Thames barge and had it moored on a permanent site at Burnham-on-Crouch. He had retired early from a high-powered job in the City for the sole purpose of living on the water. The barge was one of the most beautiful things that you would ever wish to see, particularly on the inside. This was no ordinary barge – this was the crème de la crème! Paul's parents were, without a doubt, in the money and they knew how to spend it! Most evenings there were cocktails and nibbles. Lots of laughter and hugging. Bedtime was a very relaxed affair, so we went to our bunks only when we were ready to drop. The whole of the inside of the barge was made from teak – all of it glowing with linseed oil.

Paul and I went sailing on the river in a small dinghy. There were many inlets and mud flats to circumnavigate, which were difficult to negotiate when you had no alternative but to tack. To make the task even more hazardous, you had sandbanks and very tacky dark sediment to cope with. If we strayed from the established channels, it could prove one hell of a job to get free again. We rarely saw Paul's sister as she spent most of her time at the yacht club, looking gorgeous and looking for talent.

I kept in touch with Paul for a couple of years after I had to go and live in Devon; then the war came and we lost contact, which was a great pity. Paul's parents were a delightful couple who were just mad about the sea and messing about in boats. The father worked on the barge whenever he was not sailing or entertaining. He kept the inside spotless and shining like a new pin. There was always a linseed oil rag handy. You did not require a mirror; you only had to look into one of the locker panels to see your face! The barge could sleep six with utter privacy and more if you threw caution to the wind. It had a large galley, which could cope with frequent parties, where many hands were offered to man the pumps.

Moored next to us was another barge of similar size and design to ours. It was actually tethered to us, fore and aft, and a large yacht was also tethered to them. Everyone had to come on to our deck in order to get ashore, or go by dinghy, particularly when fetching victuals. These frequent encounters made everything 'matey', especially at weekends, when most owners would generally sleep on board and make a long night of it.

They were extremely happy days and I used to look forward to my next visit immensely. I simply loved Burnham-on-Crouch; to me it was the epitome of what a nautical place should look like. It had a carefree look about it. Nothing appeared to have changed from each previous visit, and that is how it should be. It was like seeing an old friend; everything was familiar, just as we left it; a few odds and ends on the quayside... too bone idle to put them away!

It was fun to browse around in the ship chandler's. The boat fittings, polished solid brass ship's compasses, even complete binnacles as well as numerous ropes of varying sizes; some so big that they were too strong for me to put into coils. Some of the gear didn't move from one month to another. How on earth did they make a living? I am sure that the owners rarely knew the correct price. They simply looked at the buyer and if they like the cut of his jib, he got a bargain.

Unfortunately, holidays at Burnham were always over very quickly and in no time at all I was back at Woodford Green and school. But don't imagine for a moment that I didn't like Woodford Green. It was a super place and I became very fond of the whole of the Henstridge family, who looked upon me as one of their own. Luckily I had not incurred the wrath of any of the grown-ups, except for the gravel pit incident, and I have to confess to being brought home by the local bobby for throwing a stone at a greenhouse in the next road once. I have no idea what possessed me at the time.

★

Whitehall Securities had reluctantly come to the conclusion that the Spanish Civil War had turned into an international theatre of war, with long-term consequences, especially now that Germany and Italy had become involved with the Nationalists, and Russia, together with the International Brigades, made up from countries from all over the world, had become committed to an ideological struggle, with no end in sight.

As far as the company was concerned, their electricity undertaking in Almería and the province of Almería, had to be considered as lost. It was now in the hands of the Communists (Rojos to me), and they had no word from Fuerzas Motrices de Electricidad since Will Smith and Geoff had left Almería.

Geoff was given an opportunity to move to Whitehall's electricity company in Athens. This, he turned down, as he still had a long-term ambition to return to Spain and Athens might have precluded him from being considered when the time came. Instead, he accepted a position in Bideford, North Devon, as chief electrical engineer in the West of England Electricity Company, which had generating facilities in North Cornwall and North and South Devon. He decided at that point to live in the vicinity of Bideford.

In the spring of 1937, Geoff and I moved to a family hotel at Upover, Bideford, the Barnstaple road side of the river, and stayed there for a few weeks until we could find somewhere to live permanently. Bideford was, and still is, a delightful little port. The bridge built over the River Torridge is one of its main attractions. It consists of 13 arches and every one is different!

According to local legend, each arch was funded by a merchant or trader from Bideford. The size of each individual arch was in direct proportion to their wealth and the size of their contribution. It was, of course a very important port in Elizabethan times, but today it's mainly dedicated to fishing, small coastal vessels and tourism.

Geoff was very fortunate to be able to find a suitable place for us to rent. It was half of a very large house – 'Hillsborough' – between Northam and Westward Ho! It had two long drives and a huge garden on several levels. A good proportion of the grounds consisted of woodland, but there were four lawns and a tennis court too. I was over the moon with the house and the prospect of living there with Geoff.

The other part of the house was occupied by the Greathead family. There was a Mrs Greathead, very attractive and outgoing. I would say that she was in her mid-forties and there was no sign of a man in her life, but she did have a son, James, who went to Stoodleigh College as a boarder. As Mrs Greathead did not appear to have a job, we all assumed that there had to be a husband somewhere in the background.

There was a lodge at the entrance to one of the drives, which was occupied by the gardener and his wife. We had no worries with the garden as it was always kept in an immaculate state. The gardener's wife worked for the Greatheads and had nothing to do with us. We employed our own housekeeper, Mrs Smith. She lived in nearby Northam and came every morning except on Sundays. She was an excellent cook. Although she did not stay to lunch, she always left us something prepared for our supper. Mrs Smith was a very god-fearing person, as were the folk of North Devon generally. She worked very hard, did the cleaning, made the beds, and also used to take our washing home with her on a Monday and bring it back on Tuesday mornings, all beautifully ironed. My socks were also mysteriously darned!

I found out that Geoff had been engaged to be married just after the First World War, but the unfortunate girl found out that she had to compete with his Royal Enfield motorcycle and sidecar. She broke off the engagement after two years of uphill struggle and neglect. I mention this only because I became somewhat concerned when I found Mrs Greathead fluttering her eyelashes at Geoff on more than one occasion. I was anxious to keep him to myself and, in any case, my freedom would have been severely curtailed if she had had her evil way with him!

Mrs Greathead did not give up easily and had old Geoff nearly pinned to the floorboards a couple of times. Normally Geoff was pretty alert and was able to anticipate her next impending onslaught, by sidestepping out of the way just in time – but always in a most gentlemanly manner. Our housekeeper, Mrs Smith, was worth her weight in gold, as she did her very best not to give Mrs Greathead any excuse to do anything for us.

I had not been with Geoff for long before he sent me to St Boniface at Plymouth, a minor public school, as a boarder. At first I was rather unhappy because, for the first time, I was on my own as far as Spanish people were concerned. There wasn't a soul that could understand me, let alone speak to me in Spanish. My progress, or lack of it, was going to prove a real pain and, to make matters worse, I was the only foreigner at the whole school. I was one of 35 boarders and over 300 day boys.

St Boniface was run by an order of Christian brothers who ruled with a rod of iron. It was a tough regime and you either coped or you had to leave. I managed to keep going because Geoff made it very clear that there was no alternative for me. I had to learn to speak English whether I liked it or not... I had no one that could understand me or wished to; I was just a ruddy foreigner and a dago at that! Oh, how different to home, where Geoff could speak to me in my own language, and loved me. These bloody boys were so cruel, but I wasn't going to let them have the ultimate satisfaction of knowing that they really hurt me deeply.

We had two large dormitories. I was in one with 22 beds. We had no privacy and we were only a bedside table away from each other. You couldn't let your tears show or dare make any sort of noise. Most nights I fell asleep from anguish and sheer exhaustion and did not relish waking up again in the morning. Geoff came to see me and take me out at my first half-term. I told him what a miserable time I was having, but apart from a big hug, the message was plain and simple.

'Come on, I know that deep down, you have the guts to take it. You'll see… you will win!'

After Geoff's visit, I made up my mind that I was going to do whatever would help me to enjoy my time at St Boniface. I had to be accepted by the other boys, especially the leading set.

★

One of the first things to happen was an unwelcome challenge to a fight. I had to prove that I wasn't a white-livered foreigner. It was not going to be a scuffle in the playground or in some dark corner of the building… No, this was to be in the Christian brother tradition; it was to be in the gym with proper gloves! I had never boxed in my life before, but the fight was to be supervised and it soon became obvious that only fair play was going to be tolerated… thank goodness! I was a bit concerned about my opponent as he was the top bully at school and had quite a following; he was more feared than admired, but on the other hand he had never been tested!

I went down on the floor within seconds – he really did hurt! I lay on the canvas for just a moment; long enough to look up and see that I was being looked upon with derision… and that did it! I was up and possessed with injured pride. My fists went flying and I hit my opponent, McMahon, with a shower of punches to his head and body, from which he never recovered.

I was never involved in a fight again at St Boniface. I was no longer verbally abused. McMahon from then on retired into the background and even tried to become my friend, but I found it difficult to forgive him for the many public taunts and abuses that I had suffered at his hand.

The outcome of the fight was providential; it could have very easily gone the other way. It was a great boost to my decision to find other interests; it had to be a great sport, no language barriers here. I could compete on equal terms. I found rugby absolutely wonderful and I took to it like a duck to water. Football had been the only ball game that I had played before, but rugby seemed a marvellous way to release energy and develop a sense of team spirit.

In my second term, I was selected for the school Colts. I was fast and at first I played centre half. It was a superb position, for to a large extent, you were able to dictate the pace and so make the running. It also proved an excellent position for the opportune dash through the centre, by selling the odd dummy. Simply because I was unable to communicate with other boys in the early days, sport of any kind became a fetish as I found that I could excel in most of them. This new-found sporting ability earned me respect and new-found friends who were willing to put up with the language difficulties.

Unfortunately, I disappointed Geoff, term after term, by coming bottom of the class in most subjects, except in art. I was pretty good, and not quite bottom for maths, but I am afraid that overall, I came last out of a class of 32. I tried very hard to do better, but I just did not understand the teachers and they didn't understand me. As a result, Geoff became impatient and on holidays, he would no longer speak to me in Spanish. Regrettably, my school report was sent by post direct to Geoff, before we broke up for the end of term. I was therefore unable to intercept it before he could cast his eyes on it. I would have stopped at practically nothing which might have avoided him seeing such terrible results. For a few days after I got home, I kept an eye open for any letters with a Plymouth postmark, in the hope that I could intercept it, but unfortunately, I had no luck. I was now becoming a bit desperate and had a feeling that my status as a blue-eyed boy was about to come to a grinding halt!

I am not quite sure when my school report did arrive, as Geoff was in a good mood for the whole time. It was quite uncanny... I was ready for the bomb to go off at any time; then suddenly one day, which had to be several days after he had got my dreaded report, Geoff said, 'I think that you should have a real challenge, as well as a reward. If you can manage to avoid becoming last in the class next term, I will buy you a bicycle, the very best, a Raleigh Sports, how about that?'

Well, I should have been pleased I am sure, but I did not rate my chances very highly. Getting off that bottom rung of the ladder in my opinion was pretty unlikely. Geoff was somewhat disappointed in my reaction to his offer.

We had to leave things as they were and Geoff tried to make the best of a bad situation. Another term came to an end and I was back in the same predicament as before. I had given up trying to stop Geoff getting my school report, as I had become totally despondent about the whole situation and, frankly, I had for some time expected to be sent back to Spain with a flea in my ear. I felt, however, that it would prove very difficult for Geoff to organise my return, as the Civil War in Spain, if anything, was more intense than ever, so there was hope that I might have time to redeem myself.

I think that by now you have got the measure of my plight. It was a desperate situation in which I found myself... one kind word and I would have burst into tears. Trying to be brave was not easy, but I tried my best to look cheerful, even though my stomach was churning over. In the depths of despair, I saw Geoff coming towards me. He had a big grin on his face. I couldn't make it out; why was he so pleased? Ecstatic might even have been a better description. Was it possible that the school had not sent a detailed report? Had he in his generous way forgiven me for another disaster?

No. This was obviously different.

I could now see clearly that he had my report in front of him. He held it out for me to read, at the same time, pointing his finger at 'Place in Class – 30th.'

By this time Geoff was beside himself. He gave me a big hug and said 'Well done, *mijito*!' Perhaps I had done better after all. I started to get into the jollifications and the spectre of a gleaming Raleigh Sports bicycle was beginning to look like a reality. Unfortunately, I remembered that in this term, we only had 30 in the class. I had to own up. There was no alternative... I could not lie to him. I was no angel, but I valued him too much to lie. The vision of this magnificent bike was quickly receding from view.

At first Geoff appeared crestfallen, but then he came over and gave me an even bigger hug than usual and said, 'You didn't have to tell me that the numbers in the class had gone down... You shall have your new bicycle. Come with me, let's see what we have in the coal shed...'

Wow! There it was... the most wonderful bike that you have ever seen in your life!

★

Now you may find this incredible to believe, knowing the background and the situation up to now, but it eventually happened! I was no longer the last in my class. My English had improved beyond recognition and I was now able to follow, within reason, the lessons in all subjects. It happened quite suddenly. Things began to drop into place and what is more, they started to make sense.

I knew that my report for the Christmas term 1938 was going to be good. Instead of quaking in my boots, as was my usual response to the report, I was actually looking forward to my report and kept an eye out for the Plymouth postmark. I did not give Geoff any indication that I expected the results to be good, but he obviously knew that the situation was different, if only by my cheerfulness.

It was just as well that I had not given any clues, as the report when it did arrive was not quite as good as I had expected. My new place in class... fourteenth. Pretty good really, especially as the comments from the teachers concerned with each individual subject were most encouraging. This time Geoff was over the moon and kept looking at the report as if he did not quite believe it!

I suppose that fourteenth was, by comparison, hitting the dizzy heights. In some subjects, I noticed that I had come in the first half-dozen. Knowledge of English was certainly making an enormous contribution to my general success and happiness and St Boniface. I was now a long way from being just a 'wog' or a 'Spanish onion', or both. I was now better than some of my betters, as I had thought. I was also making my mark in rugby, and in sport generally.

In spite of my much-improved image and status, I found myself time and time again being jostled into leadership contests of one sort or another, for which I had great reluctance. I was encouraged by my growing band of friends and goaded by others.

I had to live down the defeat of the Spanish Armada and it was only after Dunkirk that my feelings were better understood, particularly by those who had taunted me in the past.

I became a close friend of John and Joe Groves. They were twins, in fact, identical twins, but their individual characters were very different to each other. They were lovely boys; they became so loyal to me and were always there when they were needed most. The Grove boys were in a class above me at school, but because we lived only a couple of miles away from each other, near Appledore in North Devon, we were able to stay friends. Mrs Groves was a bit like Marion had been in Woodford Green. She was very kind and used to encourage me to cycle over during the school holidays. John and Joe were quite happy to be on their own, but were more than pleased to have me join them. Mr Groves was a very quiet sort of person, but I am quite sure that he and his wife got on well.

The Greatheads left Hillsborough and shortly after, the Evans family arrived. Apart from the mother and father, who were most amiable, they had four children. The eldest, Felicity, I found to be an accomplished pianist. Then there was Francis; a very pretty girl who kept very much to herself and I saw little of her. Garry was next in line; he was a couple of years younger than me and had a mop of bright ginger hair and a face full of freckles. Finally we had Penelopy (better known to us as Pops), who was about seven years of age. She too had bright ginger hair, which was very long, and she was obviously very proud of it. She had a face full of freckles like Garry. She was a real tomboy and a lot of fun.

Geoff encouraged me to learn to play the piano and bought a grand ostensibly for that purpose. But he too was a good pianist, as well as an organist of considerable repute. I discovered much later that he had played quite regularly at the Brompton Oratory. Felicity was engaged to give me piano lessons three mornings a week during the school holidays. I wish I had taken those lessons more seriously, as I came to regret that I hadn't, later on in life.

*

Our dining room in Hillsborough was turned into a model railway room. Geoff built a most elaborate '00' railway system. It took up most of the room, where he also had his workshop. I had a small workshop next to him in what must have been a butler's pantry in the old days. The railway layout itself was extensive and it was possible to operate six trains at any one time, by using the 52-lever signal equipment. Geoff had built this interlocking comprehensive system whilst he was living in Villa Anita at Almería. It was based on the full-size principle; you could only pull the signal when the line and points were clear.

I became very keen and helped to build many of the scale model buildings at the four railway stations and also some of the countryside scenery. Geoff was a dedicated railway buff, especially on the Great Eastern and he wrote articles in the *Model Railway News* under a pseudonym. He became so well known that many people travelled down to Devon in order to see him and his by now famous model railway.

At Westward Ho! we had a fairly large beach hut. It was built by the high sea defence wall near the famous Pebble Ridge. All our friends were keen to have the use of it, as Westward Ho! beach is very open and, consequently, windy most of the time. The wind at times was more than we could cope with, as it also blew stinging sand into your eyes and face. On such occasions, we went to the lido close by, which had high walls on all sides. The other added attraction there was the café and ice cream parlour.

John and Joe often cycled with me over the Northam Burrows, where we would race on the dunes and the links golf course until the Dormy House Hotel was built and the North Devon Links Golf Course became a championship venue. Between them all, virtually the whole of the Burrows was taken over, although the sheep were allowed to continue grazing, which they do to this day, despite their droppings on the greens.

Chapter Six

During my Easter Holidays in 1938, whilst I was still suffering from my apparent underperformance cloud at school, Geoff took me to see his friend, Mrs Richmond, at Odell Manor in Bedfordshire. She was the widow of Colonel Vincent Richmond; the designer of the ill-fated airship 101, which was built at Cardington near Bedford. He was killed on the maiden flight to India, when it crashed in bad weather at Beauvais in France. Only three of the crew survived the crash.

Mrs Richmond's husband had been a school friend of Geoff's. Both went to Imperial College, London University at the same time, and they remained close friends until Vincent's death.

Odell Manor was a beautiful place and I have fond memories of my frequent visits. Mrs Richmond was most kind to me and in later years, she tried very hard to adopt me officially. She had no children of her own and lived with her sister, Nellie, who was a spinster who had lost her fiancé in the First World War. Mrs Richmond had pure white wavy hair and was called Florence, but everyone called her 'Flossy', except me. I called her Floss.

I grew to be very fond of Floss and I suppose that I looked upon her as a substitute mother and friend; definitely second to Geoff, though, in the affection tables. I had always hoped that Floss and Geoff would have married. I am quite sure that she would have been willing, if only Geoff had been a little more forthcoming. I had the feeling that this affection for her was a sort of loyalty to his dearly departed chum. They always kissed when they met, but nothing passionate.

Floss was very much the lady of the manor. In fact, her husband was to have been knighted on his return from India. The documentation and details of the coat of arms were already in hand before the departure from Cardington. Floss's family were Hodders, part of which were connected to printers and publishers Hodder and Stoughton. Her father had been a Merchant Navy captain. He also owned a lovely holiday cottage between Dawlish and Teignmouth in South Devon, called 'Minidab'. As far as I know, it is now a restaurant.

I was certainly becoming much happier at St Boniface, however, I did at first resent being a boarder, especially when I saw the day boys go home in the afternoon. I thought that they were very lucky and privileged; probably, as I thought, they did not appreciate it, having known nothing else. As time passed, the handful of boarders became more closely knit and some of us became great friends.

Generally, academic and sporting excellence were more evident with the boarders, as a percentage of the whole school, probably because of the facilities available to them in the evenings and at weekends when they had it all to themselves. I believe too, that there was a spirit of *pro bono publico* between us, which helped to emphasise and distinguish this small, elitist group from the rest.

I spent part of my summer school holiday at Hillsborough and also at Odell Manor. It was then that I met Vincent and Temperance Alston, as well as their very petite mother. Vincent was at Cranwell College, training to be a fighter pilot. Temperance lived with her mother in the village of Harrold. She was very quiet, slim and attractive in every way, as well as being pleasant, but she was always ready to slip away whenever the opportunity arose. Sir Rowland Alston, her father, had already died by the time that I first came to Odell Manor. However, I heard a great deal about him from his second wife and children. It would appear that he was a very eccentric person. For example, on one occasion, he rode his charger up the wide staircase at Odell Castle, as his new young wife had kept him waiting.

He had already been divorced when he met his second wife to be. She was a nurse and had looked after him in hospital, whilst he was recovering from an operation. Sir Rowland still required nursing at home, so she was seconded to live in at the castle for a short time, obviously long enough for love to have blossomed.

Sir Rowland had serious financial problems, despite possessing a considerable amount of property, which included a number of farms and several villages. His position eventually became so acute that he was forced to sell Odell Castle and most of his land in the near vicinity within a couple of years after his second marriage. The Alston family was of pre-Norman origin and was listed in the Doomsday Book.

The family who bought Odell Castle from Sir Rowland, the Lawson Johnsons, were the founders of the Bovril Company and had been at one time butchers of considerable repute in the nearby village of Sharnbrook. The family bought the Odell Castle estate and other property and lands. Some of the early production of Bovril was done at the Odell Mill, by the river, which was part of the original estate. Mr Johnson became a baron and took the title of Lord Luke of Pavenham.

*

When I met Floss for the first time, she was recovering from a serious abdominal operation. What exactly, I never discovered, but it took a very long time for her to feel well again. However, she was on the mend by the time I arrived on the scene. Just before her husband was killed in the R101 airship crash, he bought her an enormous Newfoundland dog as company, for when he was away at work. Wherever Floss went, Homer the dog would follow. Although he was of such huge proportions, he was as gentle as a little lamb – he was a real softy at heart and could not understand why people drew back when they saw him for the first time.

Floss and Homer were devoted to each other and he would expect to go in the pretty grey and black drophead coupé whenever she used the car. The back seat was barely big enough for him, but generally he sat up so that he could survey all before him. It became a well-known image in and around Bedford. Mrs Richmond, of course, had become a very famous person after the R101 airship crash, as a consequence of the state funeral given to the officers and crew. Their coffins were born on several gun carriages, which followed the route of the famous, along Whitehall and past the Cenotaph, and then on to the burial at the National Memorial at Cardington, in Bedfordshire. Floss, along with the other wives, followed the cortege on foot, so her photo appeared in all the nationals, the *London Illustrated News* as well as the *Graphic*.

Mrs Richmond was still grieving the loss of her husband, even as far on as 1937. I think that every time she received the very special state pension, the highest that any government had granted previously, she could not forget that this generous sum credited to her bank account every month was at the price of his life. From the letters that she had kept from him, you could not but realise that they had been deeply in love.

★

My workshop in Hillsborough served me well. Apart from the models that I had helped to make for Geoff's model railway, I also made scale model aircrafts from construction kits. My favourites were the Lysander – a high-wing monoplane used in real life for training and spotting the accuracy of field gunfire – and of course, the Spitfire, but my first model was of the Super Marine seaplane version.

A toy workshop in Mill Street, Bideford, was so pleased with the quality of workmanship and finish I achieved from their kits that they bought every model that I made. In fact, they would telephone to find out when I could make a delivery, as at times they even had a waiting list. Towards the end, they supplied me with the kits free of charge and just paid me for the make-up only. At one stage, I had little time for anything else and I was forced to restrict this money-making activity to bad weather periods during the school holidays. During term time, I was able to carry on with it after I had finished my homework. I can't say that I was rich, but definitely comfortably well off! This extra cash was welcomed as my pocket money was not adequate – only sixpence per week and it did not go up to a shilling until the start of the summer holiday in 1939.

In a way, I was somewhat fortunate to have caught ordinary measles during the summer term in 1939, as I was put into the isolation wing of the school for about four weeks. I was the only boarder to have caught it, so I was on my own for the whole period. This was to prove a very productive time. The isolation wing had its own walled garden and this gave me several walls to kick a ball against, which at times also knocked the flowers in the flower beds in front of them. The headmaster, Brother McDonald, would come and see me most days and was quite impressed with my model making activities. He felt that I was pretty lonely, which I was, and brought up a wind-up gramophone, together with some dreadful records of Wagnerian music. I still don't like Wagner, probably because I played those records so much during my spell in quarantine, as background noise.

I began to feel bad about the state of the garden and decided to start doing a bit of tidying up. Well, I was delighted to find a large, unexploded shell. I have no idea how on earth it ever got there, as you have to remember that World War Two did not commence until September 1939. This shell was about 15" long and it had a screw cap, which I could not loosen, despite many blows with a stone. I hid this shell for some time but one afternoon, the nurse walked in without my noticing her. She gave one hell of a scream and disappeared like a scalded cat. Within a couple of minutes I heard several referees' whistles being blown, which I knew meant that we were to vacate immediately, in case of fire.

Chapter Seven

Geoff and I went on holiday together for the first time that year. He had always wanted to go to Bournemouth, so that is where we went at the end of August 1939. We stayed at a small hotel nearly opposite the huge and luxurious Royal Bath Hotel. The weather was perfect and the beach was teaming with people as far as the eye could see... Not my cup of tea really. Westward Ho! was preferable to me at any time, if swimming and sunbathing were the main objectives. The town of Bournemouth was quite something else; I had never seen so many shops and restaurants. I thought the municipal band on the front was terrific.

Geoff had booked a flight to Ryde in the Isle of Wight in a twin-engine plane. This was for 3 o'clock in the afternoon on the 3rd September! We had just finished a late breakfast and were getting ready to go for a walk along the front, when we heard a news flash saying that the prime minister, Mr Chamberlain, was to make an announcement at 11.15 a.m. Everyone was expecting the worst... and they were right. We were in a state of war with Germany, as Adolf Hitler had ignored the ultimatum given to him by Great Britain, that he must pull his forces out of Poland by midnight on Saturday, 2nd September 1939.

Well I have to admit that my most immediate concern was our pending flight to the Isle of Wight. Would it be cancelled? We had no idea when hostilities would become evident. Some people certainly expected that blows would be exchanged more or less at once, probably at sea. For us, it was a sort of phoney war, but elsewhere things were beginning to happen quite quickly.

Geoff and I went to Hurn airport, more in hope than anything else. We were lucky; those who had reserved their flights in advance were to be allowed to go on the last and only flight until after the war was over. It was a great excitement. I had never been on an aircraft before. Everything seemed so small down below, when we were up in the air. When the clouds permitted, we were able to see practically the whole of the island. It was quite a small aircraft and I think that there were about six other passengers on board besides us. It was very noisy, but it appeared quite fast, as it only took about 20 minutes or so before we touched down at Ryde airport. The journey back was uneventful, although I was half expecting to see a German aircraft appear out of the skies at any time. I was a bit disappointed when it didn't!

We went home to Hillsborough the next day in our new Vauxhall. It had the latest metallic silver finish and was quite powerful, 10 horsepower. I can still remember the registration number – DUO 155; as most of our friends used to say, 'Do you owe £155?' Strangely enough, it did cost something close to that figure too!

In early October, I returned to St Boniface for the Christmas term. Everybody appeared pretty gloomy. Things had not gone well from the outset of this war. Poland had been overrun and the BEF (known as the 'Sons of the Old Contemptibles') were sent to France post haste as the German army was sweeping aside any form of resistance. To make matters worse, HMS *Courageous*, one of our largest aircraft carriers, had been sunk in the British Channel, and Russia then attacked Poland in the rear, the next day.

The aircraft carrier, *Courageous*, had been sunk by a German U-boat. These submarines were to prove a big problem for our ships bringing food and war supplies. The losses at sea were horrendous and these losses included many Royal Navy ships that were escorting the convoys across the Atlantic. We were not always told about major losses at the time that they had happened, as the powers that be were very concerned with public morale... Bad news was leaked out, often at the same time that there was something good to shout about.

Near the end of the Christmas term, we were riveted by the battle of the River Plate. It started on 13th December. The cruisers *Exeter*, *Ajax* and *Achilles* had engaged the powerful German pocket battleship, *Graf Spee*. *Graf Spee* was capable of outgunning each of the three cruisers attacking her, and because of the huge guns, she was able to fire before even the biggest of the three, HMS *Exeter*. This was our heavy, but aging, cruiser.

The ships inflicted a great deal of damage on each other, however, it was the *Graf Spee* that had to disengage and get away using her far superior speed.

She made a dash up the River Plate in Buenos Aires in the hope that she could have essential repairs carried out. Argentina was, however, neutral in the war and only allowed a short time at their dockyards for minimal repairs to make the *Graf Spee* seaworthy.

The captain of the German pocket battleship knew that the three British warships were waiting for him only three miles out to sea, and he did not know how badly damaged the HMS *Exeter* was. In particular, he had also intercepted Royal Navy signals giving details of the pending arrival of a couple of capital ships, steaming towards the mouth of the River Plate. It was, of course, very largely propaganda, as the ships in question did exist, but they were a long way away and would not have arrived in time, so the stricken German warship would have probably been able to slip past the *Exeter*, which was badly damaged. The gun ranges of the other ships were such that they would not have been able to deliver a fatal blow. The outcome was unexpected. The German captain gave every sign that he was about to make a run for it, but when out in the mid-channel on the River Plate, he scuttled his ship without any warning. As far as I know, part of it is still visible above the surface.

This famous victory was one of the best morale boosters to date. It proved that the Germans were not invincible. In February 1940, the whole of St Boniface were allowed to go to Plymouth Hoe and the centre of the city to welcome HMS *Exeter* on her triumphant return to her home port. The crew were given a tumultuous reception.

The *Exeter* had been the largest of three cruisers, but its guns and their ranges were no match to those of the *Graf Spee*. It was only the daring and courageous full steam ahead attack to get within range that won the day in the end. I believe that it was an important victory. It did more for the morale of the British people than many of the other heroic deeds that were to follow.

The older boys at school began to receive ARP training, which initially consisted of learning to use a stirrup pump; this became an indispensable piece of equipment. It was a U-shaped iron object that had a pump handle coming out of the crown part of the upside down U; one end went into a bucket full of water and the other end had a platform for your right foot, if you were right-handed! To the side you had a hose with an adjustable nozzle, which required an assistant pointing it on the desired object or area.

A tin hat was another essential part of your gear, plus a whistle and, of course, your gas mask. You went nowhere without it, not even when going to the toilet. Being a boy, requiring one hand to be free, the gas mask in the little brown cardboard box often swung in the way and got horribly wet!

I suppose that the most rewarding part of the training was to learn to enter a smoke-filled room for the purpose of looking for anyone that had succumbed to smoke inhalation. Crawling on the floor on all fours might look a bit humiliating, but I was to learn very shortly after that it could and did save lives. You can literally see clearly when your head is close to the ground.

Although you had to be 14 years of age to be an ARP deputy, I was allowed to join the first class, as my birthday was only a few weeks away. I had the good fortune to be selected to join John and Joe Groves on their firewatch duty roster. Every night, someone had to be on duty. Our watch was for alternative nights from 8 to 10 p.m. because of our age. The older you were, the later your shift.

I became very interested in aeroplane spotting and shortly joined the ATC. I was accepted by No. 197 Squadron, which was commanded by F/O Chamberlain, RAF, vr. It was good fun and I found my interest in aircraft spotting very useful. To this day, I am sure that I could spot a Heinkell 111, or a Dornier, or for that matter, a Messerschmitt 109 or Messerschmitt 110, or a Stuka dive bomber, purely by the silhouette.

By now, the Germans had bypassed the Maginot Line, simply by pushing through Belgium. By June 1940, the Germans had overrun Holland and the first bombs started to drop in Essex. Things were becoming distinctly serious; so much so that the Allies were being driven back to the sea, and the great evacuation of the BEF and Allied Forces was now inevitable... A foothold across the English Channel would now prove untenable. The only thing left to do was to salvage what we could out of the sorry mess that we found ourselves in, hence Dunkirk, at the end of May and beginning of June 1940.

The senior boys were allowed to go down to the Hoe and welcome the soldiers from the various French ports. I was now 14 and quite tall for my age, but what was more important, I felt grown-up as I was apparently looked upon as a person of responsibility. At this stage of the war, the slightest sign of initiative was pounced on. We gave cups of tea, buns and sandwiches. It wasn't that the soldiers were necessarily hungry; they were just glad to see a friendly face after their terrible ordeals. Many needed clothing, shoes and so on. Fortunately, the weather was fine and although they looked worn out, they were still able to raise a laugh and have a good butchers at any pretty girl dispensing some of the relief supplies. We weren't allowed to have cigarettes, but there were many men and women handing over whole packets, which were eagerly snapped up.

They kept coming, in a variety of boats. Some were only the size of a sloop, some quite large steamers, but the majority of the men were disembarked mainly from destroyers and frigates as they were able to dodge and fire back at the German dive bombers, the Heinkel 111 being the most fearsome machine, particularly on the largely undefended soldiers trying to clamber aboard the vessels that were able to get nearest to the shore. Of course, these were by and large privately owned, and therefore had no means to defend themselves. Those men not only had guts, they also had considerable skill to manoeuvre in congested and spontaneously organised pick-up points.

Most of the survivors that arrived where we were standing came from St Nazaire. They were a mixture of Poles, French and British Tommies. There were also quite a number of sailors of different nationalities. It was very colourful; the French had red pompoms on their caps, whilst the Poles had square hats with tassels hanging down the side. I have no idea where they all went, but they formed into a number of groups occupying the whole of the Hoe, and were gradually taken away in Army and Royal Navy transports. Some of the men did go to the Crownhill Barracks.

The Germans unleashed a deluge of bombing raids on our major cities, such as London, Bristol and of course, Plymouth. They became regular visitors. Plymouth had to rely for its defence very largely on its ack-ack mobile guns, as there were very few aircraft with which to defend the city; what there was, was not exactly state of the art – Gloucester Gladiators no less! A few were on standby at Plymouth airport. These machines were twin-wing aircraft with a single radial engine... very pretty, but nothing else!

The flying boats, the 'Short Sunderlands', were very big aircraft for the day, and were based at Plymouth. They were of constant interest to the Luftwaffe. Whenever an air raid was imminent, these graceful seaplanes would take off out of harm's way. These aircraft were based up river and were used to a great extent for submarine defence; they proved a great help against the menacing U-boats. The flying boat had a long range and therefore was ideal for coastal defence. I believe that it was designed as a bomber originally, as it had two decks. Perhaps it was the forerunner of the jumbo jet.

The raids on Plymouth were generally at night and, more often than not, during our shift. Occasionally, we had the odd daylight raid, but this was the exception rather than the rule. One day however, we were playing cricket on our best pitch in the front of the school, when a lone bomber, a Dornier 110, swooped down low, just above the school building, and fired two or three bursts of machine gun fire at our cricket pitch... This was certainly not cricket! This was our best pitch... For a foreigner you might think that it would be out of character, but you see I was now more English than the English!

Fortunately, no one was hurt. The pitch was surrounded by large fir trees beyond the cricket boundary, and those that didn't make the cover in time just threw themselves to the ground and rolled into the shape of a ball, lying on their sides. It really was a miracle that no one was hurt.

However, the pitch suffered somewhat. It was peppered with cannon shell holes that had gone far into the soft grass. It took us about an hour to fill in the divot shaped holes with bits of turf from outside the boundary line. It was a very important game as it was an inter-house match and we had to have a result! Luckily, most of the damaged turf was beyond the stumps at each end, and one bad spot right in the middle... A short ball became quite unpredictable even after using the heavy roller!

Heavy raids at night were to follow on a regular pattern, and the casualties began to get quite serious as well as the damage to poor old Plymouth. But there was still quite a look of defiance on people's faces. Nothing was as bad as it could have been – that was the attitude. Then the usual question came – how many German aircraft had we shot down? We were of the opinion that we could recognise the noise of the German aircraft; their engines sounded quite different to ours. They used to drone up and down, and you felt that they were struggling with their heavy payload; a very different sound after they had dropped their bombs!

Most of the raids were a mixture of explosive bombs and hundreds of incendiaries, some of which had explosive heads to them. We learnt to leave these for a few moments before attempting to put a sand bag over them. You could tell which were the fire and combined explosive type, as although they were the same shape and width, they were about 5" longer. They really were very nasty things, especially if they had sunk into the soft ground, as it hid the explosive part and therefore looked the same as the ordinary firebombs.

One particular night, we were all busy with a crop of fire bombs; our instructions had been to try to cover each incendiary with a small sand bag and if they got caught up in the roof rafters, we had to knock them down to the ground with a long pole, which we carried around with us. You could not really just kick them down with your foot, as they gave out tremendous heat and were difficult to look at, due to the dazzling glare. And of course, there was also the horrid smell of burning magnesium, which got up your nose.

This particular night, some of the incendiaries got caught up in the rafters and we had to clamber around the trusses. We had to be quick about it, as the incendiaries had naturally broken dozens of slates, and any aircraft above could then get a clear picture of the ground below. The incendiaries burnt very brightly and we could see clearly the field in front of the building... Only 20 yards or so away, there were about five incendiaries blazing away; they were lighting up the whole frontage of the school building.

I clambered down and picked up a sand bag on the ground near the building and rushed towards the nearest incendiary with the sand bag on my back. Suddenly I was thrown to the ground with a considerable force. I lay there for what seemed like an eternity. I waited before trying to move, first my arms and then my feet. I was all right! I was alive! I wasn't dreaming... I really wasn't hurt! I rolled over and the sand bag fell to the ground. I got up and picked up the sand bag, but in doing so I noticed that the Hessian bag had an irregular burnt shape in the middle, which would have been directly in the middle of my back. After examining it closely, I found a piece of hot, twisted shrapnel buried in the sand.

Well, it certainly had been a narrow escape. It had nothing to do with bombs, it was a piece of molten metal from an anti-aircraft shell; not unusual... What goes up must come down in some form or another!

Whenever aircraft were ahead, you could hear and sometimes see the flashes caused by the shrapnel that rained down continuously, especially from the pom-poms. These were placed around the naval dockyard in the Devonport area, and adding to the deafening noise and stench, there were shells also being fired from a mobile ack-ack battery, which had been set up just behind the school in the Plymouth Albion rugby grounds.

As you can imagine, life was far from dull. Most nights we saw some action and if not, we were always prepared for it. I don't know how true this particular incident might have been, but several of us were looking out of our first-floor classroom windows, instead of getting on with our work as per instructions from our teacher, who had had to leave us on our own for some reason or another. It was then we saw a single enemy raider, probably a Dornier 110; it was not easy to tell because of the distance and the angle at which it was flying, very low over the sea, but it soon became obvious that he was making a dive-bombing attempt on what we thought to be HMS *Newcastle*, a cruiser anchored by the breakwater alongside the old fort. Suddenly the aircraft, which was approaching the bows of the cruiser, disintegrated into a ball of fire. We felt sure that the enemy plane had been hit by a salvo of 8" shells fired from the foredeck turret of the cruiser. I have often wondered if our assumption was correct. If we were right, then it was a fantastic achievement, made possible only by the sheer audacity and initiative of the captain.

Geoff was getting very worried about the number and intensity of the air raids on Plymouth. He was able to see the red glow in the sky at night over the city, from as far away as Westward Ho! He, like other parents, telephoned to enquire about our safety; their anxiety was made worse on occasions when the telephone lines were down. The raid on 13th January 1941 proved to be the heaviest so far, and only a few days after returning to school from the Christmas holidays, when Geoff had joined the LDV (Local Defence Volunteers). With his experience and old Army rank, he was made the local commander, and I became a messenger with my new Sports Raleigh bicycle.

This particular night, I was fire watching as normal, with John and Joe Groves. Our ARP lookout was in the corridor with windows facing the school chapel. We had two walls made from sand bags and we stood between them, in order to protect ourselves from flying glass and debris from a possible near miss. On the night of 13th January, at about 9 o'clock whilst a heavy raid was in progress and bombs were falling rather closer than usual, we took cover between our protective sand bag walls, when suddenly, without warning, with no pre-swish sound, there was an almighty bang and we saw our school chapel literally disappear before our very eyes! John and I had been leaning against the wall behind us, with our feet slightly forward and together. We felt our bodies being pressed against the stone wall and at the same time our cheeks being flattened like a soft rubber ball. The hiss of air was deafening. We looked at each other and saw that we were in one piece, apart from being covered by plaster and dust. Joe had been close to us, but had been standing up. He was thrown against the stone wall behind him, which did bruise him considerably, but it hadn't been in any way life-threatening. He was able to sit down and collect his thoughts.

Within a couple of minutes, ARP from the post at the end of the main road, came over and were surprised to see us virtually unscathed. They told us that our school chapel had been hit by a land mine. These things would float down on a parachute and explode just a few feet from the ground. They were designed to cause maximum lateral damage. This explained why we had not heard it coming down.

We were left alone for quite a few nights and did not have any daytime raids for a long period either. The sudden low-flying attacks during the day had been made more difficult for the German aircraft, as the city was by now festooned with steel cables attached to barrage balloons, which were kept at a strategic height, in order to discourage low flying and to give the anti-aircraft guns a better target. These balloons looked a bit like those you see today with advertisements, but made from a proofed silver fabric.

It now seemed as if it was the other parts of the country that had their turn to get attention, especially London. During the lull periods, workmen had turned our cellars directly under the main school building into air raid shelters, which had been strengthened by 8' by 8" square wooden beams and props as supports.

★

Hell was let loose on the nights of 20th and 21st March 1941. These were by far the worst air raids so far. The centre of Plymouth and housing estates, particularly near Millbay railway station, were devastated. Everyone had to lend a hand and remove rubble in the hope of finding somebody alive. Our little group was working by the station, when we noticed that a corner of a steel table shelter was protruding through the rubble and splintered wood and masonry. We pulled everything away; some of it was extremely heavy. Soon we heard some muffled voices and then... there they were! Three people; a six- or seven-year-old girl, her mother and her grandmother. Our combined efforts proved rewarding and they could not have been more grateful. They didn't have to speak, as the relief on their faces spoke louder than words.

The steel table had saved their lives, but unless they had been released within an hour or so, they would almost certainly have suffocated. Unfortunately, a large street shelter built in the middle of a cul-de-sac nearby, suffered a direct a hit. We were moved on to deal with the huge number of incendiaries that were blazing away or just smouldering and giving off a suffocating stench. I believe that over 20,000 incendiaries were dropped over the two nights. The heat was intense, but in spite of that and the danger from collapsing buildings, the firefighters did an incredible job, particularly as the enemy aircraft lost no time or opportunity to machine-gun anyone in sight... The glow from the hundreds of individual fires lit whole areas of the city.

The school authorities, with the total agreement of the parents, decided to close the boarder section of the school, and consequently arrangements were made for the boarders to be taken home. Geoff arranged to take me and the Groves boys as we lived so close to each other. Some masters took parties to London where many lived, as it was not considered safe for them to travel on their own.

Parts of the school had been so badly damaged that large sections had to be fenced off and hence, even the day boys had to be curtailed as well. A couple of weeks after we left, most of the dayboy facilities were reopened after emergency repairs. Glass was not generally replaced; instead, a white calico type of starched material was stretched across the window frames and fixed with tin-tacks around the sides... it was easy, effective and more to the point, it gave you adequate light during the day. It also did not harm you from any bomb blasts, as glass often did. Somebody must have made a fortune!

When we arrived home to Hillsborough, we were somewhat horrified to find a fire engine at the top of the drive, pumping volumes of water into one of the windows upstairs! This was a home-from-home reception. As we got out of the car, the fireman appeared to have finished and explained to Geoff what had happened. Mrs Smith, our housekeeper, when she heard that I was coming home so suddenly, decided that my clothes needed to be aired. She had wound several items around the electric steel heating tubes in the airing cupboard. Eventually, all the clothes caught fire and then the fire spread to the whole of the bathroom and the upstairs hallway. Poor old Mrs Smith was devastated, despite Geoff's conciliatory tones as he said to her that it could have been a lot worse. Actually I was delighted; I didn't like a lot of the things that went up in smoke and I was able to choose brand new clothes and shoes; all paid for by the kindness and permission of the insurance company!

Soon after leaving St Boniface, Sir Winston Churchill went to Plymouth on a moral-boosting exercise. I was most disappointed to have missed him, as he was my hero, as he was to so many. We had to pack our things at Hillsborough and leave. The War Office had decided to commandeer Hillsborough and large sections of Westward Ho! and Northam. These were to be sealed off from ordinary members of the public, as the Americans were to use the whole area for amphibious exercises. Hillsborough was to become their officers' mess. All our furniture and books were put into storage. We went as lodgers with a very nice widow who lived at La Morna in Abbotsham Road, Bideford. The idea was for us to stay there until Geoff received his call-up papers from the Royal Navy. He had volunteered, although he was in a reserved occupation. He chose the Navy as he did not fancy ending up in the trenches again!

I was sorry to have to say goodbye to Mrs Smith, as I had become very fond of her. She had treated me like the son she had never had, and was now unlikely to at her age (mid-forties). I have a feeling that she and her husband had been unable to have children; he was a kindly sort of man and was always pleased to see me and found anything that I had to say of interest.

The LDV was changed to the Home Guard and Geoff continued to be the commander of No. 5, Bideford Platoon. As it turned out, the new headquarters were established at the Bideford grammar school. We were now living as lodgers at La Morna, which was right opposite the school. It was very convenient, as the playground in front of the school was to be for parades, marching and rifle drill.

I was quite shaken to hear Geoff give such sharp orders to the part-time soldiers, who were trying to treat the whole thing as a bit of a lark… but not for long! Geoff was no longer the kind-hearted and easy-going person that I had known for so many years. He looked very smart and I must say, whether they liked it or not, he soon had the motley collection licked into shape.

The members in the platoon were from as many walks of life as you could ever imagine: farm workers, road men, shop assistants, office workers, even the manager of the gas company, Mr Lowther, who lived in a very posh house on the other side of the road just past the grammar school. You name it, they were there. At first they tended to congregate into social classes, but soon lance corporals, corporals and sergeants drawn from different backgrounds for their leadership qualities, integrated each section into an efficient and enthusiastic team.

I well remember the day that Geoff brought home a box of live ammunition. I can't remember quite how many there were, but enough for each member of the platoon to have six rounds a piece… The war was as good as over, even with the bolt-action 303 Enfield!

Some arrangement had been made between St Boniface and Prior Park College in Bath. I believe that Geoff had been given an allowance in order that he could afford to send me to Prior Park, as it was considerably more expensive than St Boniface. I was somewhat overawed by the magnificence of the building and its sheer size; it was a quarter of a mile in length from end to end.

I had already realised that this was no ordinary school, by the uniform and accessories that poor old Geoff had to buy. You wore pinstriped trousers, waistcoats and long-tailed coats in the two winter terms. With this outfit, we had to wear stiff Eton collars, which were not only uncomfortable but lethal after they had been to the laundry a few times. They came back with serrated starched edges. The summer term was more casual, wearing our black gear on Sundays only. The rest of the week we wore grey flannels, striped blazers and cravats, all matching, and we never went without our boaters. The stripes represented the colours of your particular house.

John and Joe also joined me at Prior Park, so I had a couple of friends at least to start with. They had each other, and could therefore manage without having to make friends in the short term; however, they were very much aware that I would probably have problems getting accepted by my year group in St Paul's Senior School. The majority of the boys had come up from St Peter's Middle School and again, had probably previously to that, come up from Cricklade, the junior school, in Gloucestershire. Those boys had therefore had a long time to bond and make close friends and it was natural to them for resist intruders in the last part of their school life.

The school had obviously anticipated acceptance problems and deliberately split us up into different houses in order to integrate us as quickly as possible. Consequently, John was put into Brownlows, Joe went to Baines and I was put into Clifford. Each of us, therefore, ended up with different housemasters. Strangely enough, I fitted in quite quickly; John and Joe, however, found it difficult. Perhaps, on reflection, it wasn't so strange. In fact, I think it was cruel to separate them, as they were unhappy most of the time.

Prior Park was a world apart from St Boniface. What a toffee-nosed lot they were too… Funny how money changes people! It took a fair old time to get used to the new routine, but I did come to appreciate the system – if not the fagging tradition – where new boys became slaves to older boys.

★

Geoff had no luck with the Royal Navy, or should I say, they took too long to take him, so in sheer desperation, he volunteered for the RAF, who lost no time grabbing him because of his engineering and scientific qualifications. He was sent to Uxbridge and in a very short space of time, was made a pilot officer. This of course, meant that I would see him very rarely. At the end of term, I was told that I was going to Bedfordshire to stay with Floss. I went by train from Bath to Paddington with several boarders and a couple of masters. At Paddington, I was put in a taxi for St Pancras, where I was met by Floss, which was great. But much to my disappointment, we did not go to Odell Manor. Instead we went to the Old Rectory at Felmersham, which was the next village down the valley.

Felmersham Rectory was in fact quite large, but did not have the extensive grounds of Odell Manor. However, having said that, they were nevertheless more than adequate for my purpose and enjoyment. I met Nellie again and, on this occasion, Mabel, another sister, for the first time. She too was unmarried and very much a career type. She had recently retired as headmistress of the girls' high school at Southampton. She was now a very active writer and also reproduced classical novels in Braille.

Floss virtually became my guardian, as I saw Geoff very rarely. Whilst at Prior Park, I had all my holidays at Felmersham. We never went back to La Morna in Abbotsham Road, and I regret to say that after the war, many of our belongings had been lost. I looked forward to the holidays at Felmersham Rectory. There was a great deal of coming and going and I got to know many of Floss's friends, who, as time went on, looked upon me as her ward.

Nellie used to get quite cross at times, because she felt that Floss gave me far too much attention. She would say the same, regular as clockwork, 'You spoil that boy! His head is big enough as it is!' Actually, Nellie's bark was far worse than her bite and it was her that got me interested in oil painting. She was an extremely good artist and sold her pictures in order to augment her income.

We had no day boys at Prior Park, and hence were not unsettled as St Boniface. You would imagine that we had a lot of spare time, but in fact, we had to ration our numerous activities due to the number of clubs and hobbies available to us. Sport was as important to me as my other creative hobbies, so I had to curtail other interests, not only from a time-consuming point of view, but I also had to consider the cost of joining the various clubs, as Geoff found my school fees and the obligatory extras more than enough to cope with.

The school tried its best to avoid the less well-off from being singled out, by putting a limit on the pocket money that you were allowed to draw each week. However, my pocket money fell short of the maximum allowed. The facts were plain to see… these boys were by and large out of my bracket. One boy, for example, would often produce the odd white £5 note and you knew that there were more where that had come from! He had not handed in the surplus cash that he had brought with him at the beginning of term. His father owned a steel mill in South Wales.

I enjoyed my time at Prior Park immensely, particularly now that my English was good enough not to make it obvious that I was a foreigner. I did reasonably well at most subjects and excelled at rugby and sport generally. I was a fag for one term only as I came to Prior at the tail end of the traditional fagging period. I fagged for a sixth-form boy for this short spell and, luckily for me, he was very easy-going. He did not give me too many chores! The only thing that he would insist on was that I would bowl to him at the practice nets. He was an excellent cricketer and was in the first team. Some fags had quite a rough time, cleaning shoes, especially the rugger boots, as you had to put dubbing on after cleaning them. The system did on occasions encourage the would-be bully. I did not have to do any cleaning as we had Irish girls who made the beds and did whatever cleaning was necessary.

In my first term I shared a room with two other boys and then I had a room to myself. This was great, as I could get up to all sorts of things and of course, I did! As money was tight, I went into full production of radios with cats' whiskers. They were a bit messy to use in the dark, unless you had a torch, and earphones if you had to share a room.

I developed more powerful radios using valves and variable condensers, which I was able to pick up for a song from ex WD suppliers. The coil, which determines the wavelength, was my pièce de résistance. This was made from the cardboard core of a toilet roll on which I wound a hundred turns or so of cotton-covered solid copper wire. In order to obtain London (Radio 4 equivalent), I had to solder or tap off a lead at about the fortieth turn on the toilet roll. The exact position had to be worked out according to the diameter of the toilet roll, as I was not always able to get the same size. My interest in toilet rolls proved highly amusing to some of the masters. Little did they know that these formed the hard core of my business!

My ultra modern radio sets became an instant success; even my housemaster placed an order with me and insisted on paying the full price. Perhaps he did not wish to be beholden to me at some future disciplinary decision! As time went on, the appearance of these sets became a work of art, by my using a fret saw to cut out the front panels, behind which I bolted the small loudspeakers.

You may wonder why I had such a willing market? Well, the fact was that you could not buy radios in the shops unless you used your precious coupons. As they became more professional in appearance, I was able to charge more. The early whisker sets only commanded £2 each, but the later valve sets sold from £5 to £7 each and even at this price, they went like hot cakes. Due to the success of these radios, all the school knew about them, including the president, as he was called, so I had to hand over any cash over £10 for safekeeping until the end of term, as officially we were only allowed a maximum of £5 per week pocket money. I was able to keep £5 in order to buy fresh components for the next set. I was always anxious to avoid competition, so I kept my production methods tightly under wraps. Fortunately, such prospects were highly unlikely, as the majority of boys were so much better off that there was no burning ambition to muscle in.

I went to Felmersham for the whole of the 1941 summer holiday; going there and back on my own for the first time felt like quite an achievement, as I found everyone unable to believe that I could do things on my own at my age. My trip from Almería to Gibraltar should have been proof enough.

Floss went out of her way to make the holiday memorable. She had made certain that I would be kept busy at all times. I was now 15 years of age, but still had very little interest in girls, despite many invitations to parties given by Floss's friends, whom I am sure felt that they had to give a helping hand during the long school holidays. Perhaps they were even feeling a bit sorry for her having to cope on her own!

Floss kept herself pretty busy too, what with her various social and voluntary commitments. She was, for example, head of the WVS for her area in Bedford; she was a governor of Carlton School and endless other committees, which sometimes included Captain Starey. He lived in a magnificent house, Milton Hall in Milton Ernest. A close neighbour was Sir Richard Wells, the boss of the Wells and Winch Brewery, who would accompany her to some function or other and lived at the Grange, close by.

During the summer, Temperance Alston came to live with us as her mother had died rather suddenly and she was left high and dry with nowhere to go. She was an extremely attractive girl and consequently she was paid a considerable amount of attention, particularly by Russell, the Duke of Bedford's eldest son. She could not have been less interested and made no bones about this, not trying to hide the fact. I had to act as a go-between, as Temperance rather hoped to cool his ardour by having no direct contact with him.

He would often call without warning, in his father's chauffeur-driven Rolls, in the hope of catching Temperance off her guard. But this was to no avail as, invariably, I had to tell him that she was either too busy or that she had a migraine and could not possibly see him. The migraine excuse was often true, for she really suffered from them badly. After a few months of being so rebuffed, Russell accepted their family motto *qué será, será*.

Temperance joined the Women's Land Army in anticipation of being called up by one of the services. This was a very sought-after wartime outlet for the young, 'county-class' females, who were not attracted to the WRNS, the only other, suitably genteel job for a young lady! Temperance certainly landed herself a very cushy occupation, working for Sir Richard Wells on his estate. Floss, no doubt, also had something to do with it.

Temperance looked very smart in her Land Army uniform and perhaps for the first time, she awoke some feelings in me; up to now girls had been part of the everyday scenery and passed me by, but she was very pretty and was constantly noticed by others, young and not-so-young. So I was only reacting like any other red-blooded individual, and that was as far as it went.

The house was overflowing, as Floss wanted to do her bit for the war effort and took in four evacuees from the East End of London, two girls and two boys. Floss had no intention of having so many, but when we arrived at Sharnbrook railway station to pick them up, she could not bear to see the faces of these four young brothers and sisters when the organisers tried to allocate them over more than one family. Their sad faces drew nothing but compassion from everyone around. Floss Richmond had the biggest house, so in no time at all was the proud possessor of four extremely happy young evacuees.

Our freshly acquired family were over the moon to be going with us to a new life in the country. They had never been outside London before and everything was exciting; they had never seen farm animals close up before, only in pictures. Kathy was the eldest, I think she was about 14, going on 15. She was to help in the house and assist Sarah, the cook. Sarah was getting on a bit and had been with Floss from the day she married Colonel Richmond. We had seven main bedrooms as well as two attic rooms which, years ago, had been for the servants. Kathy and her sister shared a room and the two boys shared another, so when Geoff came home on leave, I had to sleep on the sofa in the study/library.

That summer, life became very hectic and the war was, by default, put on the back-burner, although it had spread even further; Hitler attacked Russia in June... perhaps a good thing! We all felt sure that Russia could take a lot of the sting from Hitler's tail and consequently give us breathing space. However, at first Germany swept everything before it; it was not until Stalingrad and the Kiev front that Hitler began to see that he had made a terrible blunder in taking on Russia before trying to finish us off first.

I went back to Prior Park in September 1941, and found that the Admiralty had taken over the new wing, which contained our new theatre and up-to-date laboratory, as well as classrooms for the upper fifths, first-year sixths and the very latest in gymnasiums. We had been waiting for our new facilities for so long that the occupation of them by the Admiralty was a bit of a surprise to say the least.

We were kept firmly away from the new wing by armed guards with fixed bayonets. There was a lot of coming and going of RN vehicles, which brought officers and personnel to and fro, some with a considerable amount of 'scrambled egg' on their caps – military decorations that started at rear admiral in the Navy and lieutenant general in the Army... So this was obviously to be an important naval base. Some of the parents, I recollect, were not too happy to find a military establishment so close to our living quarters, particularly after they found out that it was a naval intelligence base. We, on the contrary, found it quite exciting and never gave any thought to any possible danger!

I was on cloud nine, so to speak, as I had been selected to play rugger for the first team. I played either stand-off or, more often than not, at number eleven in the three quarters. Brother Eric

Burke was the senior games master – he had at one time played for the Irish Republic. The deputy games master, Brother Monahan occasionally still played for southern Ireland.

We could hear the German bombers flying over to Bristol on many nights and thanked our lucky stars that it was not to be our turn this time. Soon we would see the glow in the sky which would inevitably follow these raids. I was extremely upset to learn from a phone call from Geoff that his sister Marion had been killed by an explosive incendiary bomb whilst staying in Bristol, the sort of bomb that I remembered only too well.

It appeared that Marion had left London with her two youngest children, Vincent and Sheila, in order to get away from the heavy bombing and disturbed nights. They had been staying in a small family hotel quite near the centre of Bristol. Someone had obviously given her very poor advice, as Bristol was no different to London, just on a smaller scale. The explosive incendiary had landed on a bay window roof and as she found it, it was burning and melting the lead covering. She had opened the sash window of her bedroom to step out on the roof and try to push it off when it blew up, killing her instantly.

Chapter Eight

Prior Park was having one if its better years at rugby. We won all seven of our fixtures; in fact, the first team had achieved such a high standard that we received an invitation to play against a grown-up side, none other than Hayden Tanner's Welsh! The team was made up from servicemen on leave, but who also regularly played for their own service rugby teams. We lost 5-8, but this was still a good result, taking everything into consideration. Brother Burke was so pleased that he arranged a match with the Bath Rugby Club. Fortunately I cannot remember the score, but it was a bit of a disaster!

On the news, we heard that Japan had bombed Pearl Harbour, sinking most of the American Pacific Fleet. It came as no surprise when America declared war on Japan, as well as Germany, at the same time as us, on 8th December 1941. Soon after, a commando unit encamped on our cricket pitch early one morning, for which we, the boys, had no prior warning. We had never seen these things before, but they had little foldaway miniature motorised scooters. I believe that they were called 'corgis' – absolutely brilliant! We tried to ride them, but there were too many officers about. I was hoping that our cadet force would be allowed to go on manoeuvres with them, but they did not stay long, and went as quickly as they came.

I was again off to Felmersham for the Christmas holiday. I really looked forward to seeing Floss as she was such fun and, I had to admit, also a bit of a snob, but she got away with it! Vincent Alston came to see his sister Temperance and brought his friend Johnny Johnson with him. Both of them were Battle of Britain fighter pilots and Johnny was by then well known for his dogfight achievements, but always looked as if he could have done with a good night's sleep. He was a most likeable person; he usually had a mop of unruly blond hair and a wicked but engaging smile on his face. I noticed that he was quite keen on Temperance, so probably that was one of the main reasons for his visits with Vincent, but both were friends of Captain McCallah, who was the son of the major who lived a couple of miles away. The major had been chief of police in Lagos, Nigeria. He spoke with a typical staccato voice expected from a man who had held those sorts of positions. You could hear him from a mile off!

The major was a frequent visitor for morning coffee and always carried a cane under his arm wherever he went. As you would imagine, his wife was every bit the quiet, dutiful person that he would expect and demand. In spite of his brusque manner, he was all right deep down. In fact, in his own mind, he considered it his duty to keep an eye on Floss, as he looked upon her as a defenceless woman needing protection.

On one of the major's coffee morning visits, he brought with him his niece, Greer Garson, the well-known film actress. She was a charming person and we enjoyed meeting her as she was not only pretty but also very vivacious. She must have got her looks from her aunt's side of the family. It couldn't have been from the major!

During the holiday, I made myself a workshop in the garage and bought a few essential tools from Woolworths; nothing cost more than sixpence (2½ pence, new money). If the article did cost more, then it was sold dismantled, and each piece priced separately. Repairs were mounting up and Floss couldn't get anyone to do any sort of maintenance. The only fellow who would come, at a push, was the undertaker in the village. He presumably was on essential war work and so not called up – come to think of it, I am sure that he was in his fifties! He was very proud of his coffins and always had some on display in his sitting-room window.

Geoff had put me through a rigorous workshop practice. This was now to prove of considerable value, particularly as Felmersham Rectory was extremely old and required a continual maintenance programme. The state of my finances would not have allowed me to get every tool that I would need, so a kind village undertaker was an essential ally who would allow me to use his facilities or to borrow some of the more specialised equipment.

George's undertaking business took up all of his available working hours, so he was happy to oblige me and hoped that by doing so, my wide connections would put him in a favourable position when his undertaking expertise might be required. It was quite eerie when I first started to use his workshop, as it was next to the room where he laid out his customers before being put into a coffin. The coffins were generally made to measure and then wheeled into the chapel of rest, as he would call it.

One afternoon, soon after finishing lunch, we were sipping a cup of coffee by the French window in the dining room, enjoying the view up the garden on a lovely sunny April day, when the front doorbell rang with a startling harsh sound that shattered our peace and tranquillity. The bell consisted of a big solid brass dome, which you had to twist around in order to wind up the clockwork mechanism behind the door… it had been fully wound, hence the sharp sound that had surprised us. Callers rarely used this bell as the front door was generally open, weather permitting, and visitors then rang the electric bell on the inside door of the outer hall, or the pull-bell brass handle which would ring in the kitchen and activate one of the numbers on the display board over the kitchen door. Friends frankly did not ring any bell… they just walked in and said 'Cooee' or words to that effect!

Who on earth could this be? It certainly could not be any of our friends, as they knew the drill. Well, somebody had to go… All eyes were on me. I got up rather hesitantly and went to the front door with air of expectancy and puzzlement… Kathy was probably having her late lunch, and in any case would not have heard the bell in the servants' living room/dining room.

I was somewhat taken aback and surprised, for in front of me was a very good-looking young man with piercing blue eyes, standing as if to attention.

'Have you any knives, scissors or shears that need sharpening?' he said this in a clipped manner without offering the usual opening greeting. I was amazed to say the least, as he was, first of all, of obvious military age and looked pretty fit, so why was he not in uniform? Secondly, his near perfect English accent, educated to the point of refinement, could never have belonged to a grinder or tinker; that is if they even existed at this period of the war!

I said no, as I do those sorts of jobs in the house and asked, 'By the way, where is your machine?'

'Ho, my brother has it just up the road,' he said. This meant that there were now two unlikely characters. He must have realised that things were going sadly wrong and quickly touched his cap and turned on his heels.

Floss reported the incident to the nearest constabulary, which was at Sharnbrook. The police there were most attentive – they were positively excited. It appeared that a couple of days earlier they had picked up a German saboteur on the bridge over the railway line at Sharnbrook. This was not only the main line to Scotland, but also the express section of the LMS from St Pancras in London.

A person fitting this description was seen to board a bus at the village crossroads, which was bound for Bedford. He was arrested in the terminal in Bedford. Yes indeed, our man was a German spy! They thanked us for our vigilance and said that he had been sizing up the place in order to install a transmitter in one of our outbuildings.

I was due to return to Prior Park; this would have been at the end of April 1942, when I received a phone call instructing me not to return for another week. Prior Park had been bombed and had been hit by two sticks of four high-explosive bombs. St Paul's wing had been gutted and our new block occupied by the Admiralty had also been badly damaged. This was the time when Hitler decided that the cathedral and historical cities should be reduced to rubble in order to demoralise the public at large. Needless to say, it did not work, in spite of the devastation which it caused in cities such as Exeter, Norwich, Bath and York, where much of our heritage was destroyed. Prior Park was obviously in that category, however, I do believe that the Germans knew of the significance of Prior Park as a naval intelligence base and that was the reason for the accuracy of their bombing of the school.

It may sound callous, but at the time all I thought about was the fact that we had got another week's holiday and that was something else – hurray! On the way back on the train, it began to dawn on me that the situation was serious. Had any of the masters been hurt? I did like my school a lot, it was a beautiful place, so I was anxious to see how much it had suffered. Today people pay to see it. The glorious gardens, the priory and the palladium bridge, down by the lake, were magnificent. I believe that there are only three such bridges in the country.

I was quite glad to get back and hoped that perhaps things were not going to be as bad as first feared. The memories of the damage that I had seen at Plymouth made me apprehensive. Well, it was quite bad. Nothing much appeared to have been done since the raid on Bath; certainly not by the railway station where the damage had been considerable. I didn't see the centre as I got a taxi and went straight up to Prior Park Road. First impression on arrival at the school was that it was a total shambles. Debris was everywhere, mountains of it. Broken sections of beautifully carved sandstone architraves were lying about where they had fallen.

The school was built of the lovely soft Bath sand-coloured stone, as of course was most of Bath. My bedroom had suffered a direct hit, along with several others. The St Paul's wing of the school was gutted so we had to be rehoused in the Mansion House, the middle part of the whole building. We were to sleep in the banqueting hall, which had been turned into a makeshift dormitory. I found my bed directly under one of huge chandeliers!

I was not exactly thrilled at the thought of this thing hanging over me like the sword of Damocles, especially if we were to have more raids! However, there was no way that I could complain after having seen poor old Brother Burke coming towards me, hobbling on a stick. He told me that at the time of the raid he was at the top of the winding marble staircase in the St Paul's Building. He was thrown to the first floor landing. He was bruised, but, miraculously, nothing had been broken.

★

It was the term for playing hockey. I found it nearly as enjoyable as rugger. In previous years, before my time, Prior Park had been top of the public school sides in the country. In fact, they went on a tour of India. We still had a good side, but wartime restrictions made it necessary to keep school fixtures closer to home, so we did not get such a wide experience. Consequently we were not in a position to judge the standard achieved by other schools.

Prior Park was at this time very advanced with its sporting facilities; for example, we had an international-size indoor heated swimming pool. We had the feeling that nothing was too expensive if it added to the kudos and reputation of the school. Prior Park gave me a great insight into the British character. It was not only complex, but it was also not always direct, so much so that at times I felt rather naïve. A direct question rarely received a direct reply. It was generally couched in some form or another. Perhaps this is a public school phenomenon; for example, all pubs were out of bounds, yet we had a superb pub in the middle of Combe Down Village. Not only did the landlord make us very welcome there, but the pub had a Prior Park Room sign written over the door of the snuggery!

It was not on to tell your housemaster or the president, Brother Robinson, as he was called; he would have already known of the existence of this room in the pub as it was specifically reserved for Prior Park boys only. Officially, however, it did not exist. The head relied on the good sense of the landlord to avoid any public spectacle. The criteria for most offences was simple… don't get caught, but if you should be so unfortunate, you suffer the consequences without a murmur. To an outsider, this would have appeared as unfair or, at the very least, obtuse. To be a foreigner, nay, a wog, would make it impossible to understand, unless you had had prior training in such matters. You are probably wondering how I was able to detect this peculiar characteristic… Well, you see, I had a head start, as my great-great-great grandmother was an American, Sarah Anne Mills, and what's more, she was of West Country descent.

The hockey term went off rather uneventfully. The Admiralty moved out and left a trail of documents in the drawers of the desks which had been in use – some were even rubber-stamped 'Top Secret'. We were not able to make them out, as they were in code. The Admiralty was obviously able to pull strings, as builders came within days to make emergency repairs. The St Paul's wing was too badly damaged to be repaired at that moment, and in any case, would have required skilled stonemasons. We did not, therefore reoccupy it in my time and consequently I never went back to my old bedroom.

Queen Mary often visited the school, when she was invariably accompanied by the two princesses, Elizabeth and Margaret. They would come in an old green Daimler, which became well known as the 'barouche'. At the time, Queen Mary was living at Badminton, which was only a stone's throw away from us. I believe that she stayed there for the best part of the war. She was a great admirer and collector of antiques. It was not unusual for her to pass comment if she was particularly interested and keen to possess a particular article. It was also quite likely that it would be awaiting collection on her next visit.

The Mansion House was full of antiques and wonderful paintings. It had been the home of Ralph Allen, who had been given the rights to the crossposts linking Bath, the West Country and the South-west to London. Ralph Allen apparently earned £16,000 from the above franchise in the first year alone; this, together with the money made from quarrying and shipping Bath stone to every corner of the country, enabled him to build Prior Park between 1735 and 1742. He became very influential and such people as the poet Alexander Pope, actors Quinn and Garrick, as well as Pitt the Elder, the prime minister and the MP for Bath, were regular visitors to Prior Park. Even Princess Amelia, the sister of the Duke of York, resided there for a period of time as a guest.

The college was founded by Bishop Baine in 1830, I believe with monks and students from Ampleforth Abbey. Relations between Downside Abbey, Prior Park and Ampleforth were not particularly happy ones, especially as there had been moves to close Downside and preference had been given to Prior Park College. Downside Abbey still has possession of the Prior Park monstrance, which is made from solid gold and studded with jewels. The feud continues!

On an April afternoon, Queen Mary, unbeknownst to anyone, arrived in her green Daimler with, as was now usual, the two princesses. But instead of coming along the top drive and past the lodge, she had driven up the very long lower drive by the Priory Grange at the entrance to the Prior Park grounds. She immediately went up the winding front stone stairs to the door of the Mansion House overlooking the city of Bath... finding it full of boys in different stages of undress! She retreated smartly back down again and went through a doorway leading to St Paul's where we were still able to use the shower rooms.

We had just finished a game of hockey and I was rushing back to the dormitory-cum-banqueting hall/museum when wham! I had my head down and had hit something pretty hard. I looked up from the feet upwards to find that I literally had rammed a lady with a blue coat... This could only be her! I heard her say, 'I do believe that it hurt you more than it did me!' tapping her whaleboned stomach as she spoke.

By this time my housemaster had arrived on the scene and I knew, to put it mildly, that I was 'for it'. Not only was I using the short cut, but I had also been running... Both mortal sins! Queen Mary was most magnanimous and diffused the whole thing by saying, 'Go on your way; it was obviously very urgent.'

I departed, bowing and running backwards at the same time, until I was around the corner and out of sight. Nothing more was said about the incident, except that this particular doorway was, after that, firmly closed.

★

I was looking forward to the holiday at Felmersham as I missed Floss. She was such fun. I crossed off each week on my calendar and, sure enough, there I was, back again. I met Dr Liddell and his twin brother that holiday. They were both Harley Street specialists. Both were unmarried and great friends of the journalist, Ward Price. I found them interesting and invigorating as they brought with them the latest thinking and social development evolving in the metropolis and, therefore, kept us up-to-date in our backwater. Their regard for my views as a young man was flattering, to say the least. They made me feel special by wishing to spend so much valuable time chatting to me.

Miss Day came from morning coffee quite frequently. She was a very proficient watercolour artist. One morning, she did one of her paintings whilst sitting having coffee. It was a magnificent view of the garden from the dining-room French windows, which opened out onto a lovely part of the lower garden. Coffee time always made me nervous, as Floss insisted on drinking out of her very delicate Minton coffee cups. I learnt a great deal about fine china and silver as well as antiques. The vicarage, as we preferred to call it, was full of lovely things, including a wonderful collection of oil paintings.

In the sitting room, on a solid carved double-pedestal oak desk, were a pair of exquisite Meissen Dresden china candelabras with supporting angel candle holders. Some years later Floss gave the desk to Geoff, who had always admired it. It went from St Leonard's in Exeter to La Rosa in Silverton, but ultimately, when we moved to Uffculme, it was regrettably sold, due to the lack of space.

The paintings included many from famous artists such as Birleigh Bruell and a watercolour by Sir William Russell Flint. The latter had been a friend and colleague of Floss's husband, Colonel Richmond, when he was building the R101 airship at Cardington. Both were keen on amateur dramatics and Russell's job was to deal with the scenery, especially backdrops, for the stage productions. Floss gave me a little gem of a watercolour that Russell had painted and sent to Dopy (Vincent Richmond). It was his suggested idea for a backcloth to be included for the stage show at Cardington. On the back of the painting, Russell had written, 'I hope that you like this suggested backcloth. Sorry Dopy. I know that the flowers depicted do not all bloom at the same time, but I thought that they were effective!'

He called Colonel Richmond 'Dopy' because he invented the proofing solution for coating the pure Irish linen gasbags used in the R101 airship. Very sadly, this little treasure of a painting has either been stolen or mislaid in one of our house moves. I would like it back, so I hope that somebody will have the compassion and take pity on me by returning it one day.

I liked Russell, but at that time, when I was in Bedford, he was only just getting to be known. His watercolours were brilliant, especially one of his Spanish wife in the nude. The washerwomen scrubbing clothes on stone slabs in an outdoor washhouse has since, of course, become very famous.

It was good fun being in the Mansion House, as we were so many in one room. Pillow fights were a regular event and were high up on the menu. Picking up feathers in the morning wasn't quite such fun, as by then the doors and windows had been opened and loose fluff and feathers had found their way into every nook and cranny – very difficult without a vacuum cleaner!

One night when the lights were put out so that the doors and windows could be opened, someone had hurled a hockey stick across the room in a fit of temper and hit the ornate gilt frame of one of the most valuable paintings in the banqueting hall, none other than The Charge of the Light Brigade at the Battle of Balaklava. It could have been much worse; it only took off a small chip of gilt from the frame. We knew where to look, but no one else would ever find it.

The food was generally foul. I think that was the only way that it could be described. Complaints were numerous and even the prefects agreed and were prepared to support us. The stock answer was always the same, 'Haven't you heard? There's a war on.'

I do believe that the school did buy good quality food; it's just what they did to it that made it so disgusting. The best meal by far was breakfast on a Saturday, as this consisted of a large pork sausage and a rasher of bacon. On Sundays, we had two doorsteps of fried bread and a great dollop of thick-cut marmalade on the top. Strange as it may seem, it was really delicious. And still is!

At night we took it in turns to raid the kitchens. Whoever was in charge that night took the orders for toast, sandwiches and baked beans. We would open at least one large golden tin of the real American Spam, which was quite different to the rubbish sold after the war under the same name and made in this country. If you never had the real thing, you haven't lived! You couldn't mistake these tins as they were about 12" long and 4" x 4" square and, of course, light gold. They were sent to Britain under the lease-lend system.

The sandwiches had to be seen to be believed, as they were more like hamburgers are today. To make life easy for the catering staff, the school ordered thick-sliced or uncut bread only. We never appeared to be short of butter, like most people were. You might wonder why we were not caught stealing so much food, especially as it was meant to be strictly rationed. There was a very simple explanation for that; you see, the establishment was always able to claim that there were more people at the college at any one time with so many comings and goings. No one could ever prove otherwise. Shortages never came to light as so many people were on the fiddle, from kitchen personnel to senior staff. In fact, we did not have to keep quiet during our raids to the kitchen as the others kept a lookout in order to see when the coast was clear for their turn!

It was quite natural to get up to mischief, particularly with so many of us in one room. A dare of some sort or another at least once a week was inevitable. One night, a group of boys climbed up the bell tower and rang the large bronze bell. It caused pandemonium. The sounding of church bells was to act as an alarm for either a gas attack or as a warning of an imminent invasion by the Germans. The ARP wardens came up from the city post haste to enquire if the alarm was genuine! The school received a formal complaint and the head took it very seriously. The culprits were never caught, and consequently we were all made to suffer by the loss of privileges for a whole week.

Unfortunately, I was caught one night, right in the middle of the act. I had agreed to go into the main chapel and unscrew the lid of a coffin, which had been left overnight on trestles in the side chapel. I had managed to remove about three-quarters of the wood screws when the smell and fear of seeing a mutilated body put me off my dare. The body was that of an Irish workman who had been run over by a lorry, which reversed while he had been at the back leaning on his shovel.

In my haste to put back the screws, I damaged the dome heads which covered them. By now, I was full of remorse. The thought that this unfortunate Irishman had been somebody's loved one made me a bit unsteady as I rushed through the fanlight and walked between the plastered and embossed domed ceiling and main roof rafters. In my haste and excitement, my foot slipped off one of the wooden beams and poked through the ornate moulded ceiling.

Plaster went crashing down with a deafening noise, breaking the stillness of the night and of all places, directly over the main altar. The deafening clatter disturbed the janitor and he unlocked the church door, saw the mess and caught me as I let myself down the fanlight at the other end of the ceiling.

For the episode, I was very nearly expelled and I am sure that I would have been, if it had not have been for Brother Burke, who interceded on my behalf. I think that he was persuaded by his concern for the First rugger team; I had just recently been awarded my colours. However, it was decided that Geoff should be told of my escapade. He took it very well. School reports were now much more to his liking, but nevertheless, I got one of his rare but severe dressings-down. He made it abundantly clear that I was not to endanger my place at Prior Park, especially as he was having to make considerable sacrifices to keep me there.

★

When I went home for Christmas, Floss suggested that I should dress up as Father Christmas and fill the stockings of the two young boys, who still believed in him. I think they were having me on, to ensure they got the presents they wanted! Geoff arrived on Christmas Eve and had quite a few days' leave. He was very jolly as he had just got a further promotion; he was now a flight lieutenant. I was, of course, very proud of him, but on the other hand, I had expected nothing less of him as I held him on a pedestal and therefore believed he was capable of being the best at anything that he did. His ability knew no bounds.

We decided to see the New Year in, hoping that 1943 was going to prove to be the year that our fortunes changed for the better. We – Geoff, Floss and I – were finishing our drinks on the patio outside the dining room window, when Geoff remarked that he could see sparks flying up into the air just to the left of the village church. We both went out into the main road and could now see clearly that it was not the church that was alight, but a cottage a little further down the road.

By the time we had got to it, the roof was well alight. The smoke, heat and crackling had awoken the family who occupied the cottage, and they were already outside, watching their cottage being devoured by flames, which were now several feet above the thatched roof. They stood there, wrapped in blankets over their nightclothes, eyes wide open and completely silent.

It was a bitterly cold night and frost had already settled on the tarmac surface of the road, which now glistened from the light of the flames. We all knew the family well; the father was a farmhand at Hensman Farm just up the road, and the mother worked for the local doctor. They had two daughters; one was a teenager and the other, Jill, was only about seven years of age.

Geoff went off immediately to phone the fire brigade from the call box beside the church lych gate. Suddenly, Jill rushed back into the cottage. Fortunately I had nothing stopping me, so I was able to follow her in within seconds by crawling on all fours. The heat was bad enough, but the dense smoke billowing down the wide staircase was a frightening deterrent. I scrambled up partway and was about to give up my search, when I saw Jill lying face down, with her head pointing towards the ground floor, near the top of the now partially charred stairs. She was motionless… but clutching her precious dolly.

Jill's mother was horror-stricken when she saw me carrying her little girl, and grabbed Jill from me. It was a miraculous escape; only her hair had been slightly singed. Within no time at all she began to cough and splutter and her mother's relief was such that she very nearly instinctively smacked her for being such a silly girl!

*

The Christmas holidays were soon over and I was back at school and not feeling very happy as my best friends, John and Joe Groves, had left at the end of the previous term. They couldn't wait to go and join the Merchant Navy as they had hated their time at Prior Park. It was but a very short time before Joe was lost at sea, on a convoy to Murmansk in northern Russia. John had a hair-raising experience on the famous convoy to Malta, which brought vital supplies for the relief and defence of the island. For the heroic stand and bravery of all the inhabitants of Malta, they were collectively awarded the George Cross. After the war, John became the harbour master at Pinang (George Town) in Malaysia and married Vicki, a white Russian.

This particular term at school proved uneventful for me, but Geoff was having problems with my family in Spain, to be precise, with Tío Manuel. The Civil War had come to an end a few months before the start of the Second World War. Tío Manuel had been set free from prison by the Franco forces. He was able to return to the electricity company and was made managing director. He was also appointed MD of a chemical factory near Almería, in which he had a large shareholding. I had received a letter from Tío Manuel a few weeks earlier, asking me to return to Spain without delay. He told me that he had missed me and thought a lot about me whilst he was held prisoner in the hold of an old cargo ship in Almería harbour. Marisol also enclosed a translation of his letter, in case I had forgotten how to speak Castilian Spanish. I replied promptly saying that I did not wish to return to Spain as I had made England my home.

About three weeks later, Geoff was charged by my uncle, in a London court, to the effect that I had been abducted by Mr Websdale against my wishes and prevented from returning to Spain after the Civil War.

Geoff was very upset, as there was no vestige of truth in the allegations. After considerable pressure from the Spanish embassy in London, the court left it to me to make a voluntary decision without pressure from either party. If my decision had been to return to Spain, then a seat would have been reserved on the Lisbon plane, as there were no direct flights from London to any part of Spain.

Geoff should never have had any doubt as to my decision. Not only did I think the world of him, but I was also grateful for all that he had done for me, let alone the considerable sacrifice that he had made on my behalf by sending me to a good boarding school, which he could ill afford. The court case made Geoff realise that Tío Manuel must have had considerable clout to be able to bring a court action at the height of the Second World War.

He warned me never to divulge my relationship to Serrano Suner, General Franco's brother-in-law. Not only was he my uncle by marriage, but he was also secretary general of the *Falanje*, the Spanish Fascist party, and on fairly close terms with Adolf

Hitler, Ribbentrop and Mussolini. At the time, I was in no position to argue with him, even though I felt that he had been misinformed. But I wasn't going to take any chances as I could have been interned.

Chapter Nine

I was now a senior and in the upper fifth. From now on we were allowed ale with our supper. It was pretty weak stuff, but very welcome nevertheless. We had considerably more freedom as seniors and did not have to account for every minute of the day. It was not unnatural for boys at this age to turn their minds to thinking of girls. It was not unheard of for an Irish girl to be sent back to Ireland before it became obvious that things were not exactly as they should have been! We all thought that this was a very dangerous arrangement for these skivvies, as we used to call them, to keep our bedrooms tidy. The majority of these girls were most attractive, with their blue eyes, dark hair and Irish brogue. It was a recipe for temptation and impending disaster to employ them to polish the floors of our bedrooms, clean windows and make the beds. This was more than the premature shaver boys could resist. A romp on the bed before making them up was a good sport at any time, and what is more, it was usually welcomed!

The move to the Mansion House meant that some of the four o'clock shadow brigade were not getting their oats, so they had to look for pastures new, such as the Connies of this world up in the village. They were game if any of the more mature boys wanted a tussle or two. Quite a number of the so-called conquests had to be taken with a pinch of salt, as much was bravado and in fact, many had little idea of what it was all about! Certainly not judging from some of their lurid accounts.

Once again, I was on my way to Felmersham. I came to love that place as much as Odell Manor, although it was not in the same category historically, in size or prestige. It was not a bit imposing. However, I was to discover that it had quite a history. Martin, the vicar's son, had become a very good friend to me and loaned me a book about the village and, in particular, Felmersham

Vicarage, which was supposed to have a famous ghost. Well, I was quite sure that the ghost did exist, from some of the experiences I had encountered there.

After reading the book, I went in search of the priest's hole as well. Sure enough, after careful checking and rechecking, I found a gap between the outer wall and the fireplace in Nellie's bedroom. The chimney had two cupboards on either side, one above the other, and each had a pair of doors, with the top cupboard being much smaller. The cupboards on the left-hand side, next to the outside wall were only two feet in depth, leaving about another four feet to the end of the chimney breast. The priest's hole had to be behind this particular cupboard.

I got a stepladder and climbed into the top left-hand cupboard to find that someone had papered the three sides. However, I noticed that there was a slight upside down U-shaped ridge showing behind the wallpaper, just above the shelf. This had to be a door, ever so small, but definitely a door. With a sharp knife, I cut that wallpaper around and found that it was a wooden panel with a pair of hinges, which I was able to prise open.

I was apprehensive and full of excitement as I expected to be overcome by the strong musty smell and to then find a partly decomposed body, or at the very least a skeleton with a dog collar around its neck. But nothing, not even a smell!

A wall ladder ran down to the ground level but there were two ledges, each about two and a half feet wide, part-way. There had to be a vent somewhere as the air was pretty fresh. It turned out to be a very simple solution. At each floor level there was a gap left in the wall and I was able to see that it became a duct between the floorboards and the ceiling. These in turn had a crude vent to the outside wall neatly behind the downpipe of the guttering, of which there were three, causing a sufficient flow of fresh air to pass through the cavity.

The priest's hole was a bit of an anticlimax in a way, as I did not find anything gruesome; in fact, I found nothing at all. I have a feeling that this had been discovered long ago. Perhaps some poor chap had been caught... We will never know! The missing space between the study/library and the servants' sitting room that I discovered had been very cleverly disguised, so I was quite chuffed about it really, and it did also give me something to talk about with our regular visitors.

Although there had not been the slightest hint of a musty smell, Nellie was convinced that she could now detect it. Frankly, she was never happy with her bedroom again and was so uncomfortable with it, she even tried to change over with me, particularly as she considered it to be my fault!

★

Captain Starey introduced Floss to several American officers who were based at Milton Ernest. They had taken over Milton Hall and the entire estate. The Americans had built dozens of Nissen huts dispersed amongst the trees surrounding the estate, which gave good cover from the air. The Americans generously allowed Captain Starey to live in a small part of his property and treated him with tremendous deference and generosity with the food that they supplied him from their officers' mess.

The Americans were all over the place. First of all we had the eight American army airfields; the largest and nearest being the one at Thurleigh, where the flying fortresses were based and from where the huge daylight raids over Germany were largely organised. Then we had the massive concentration of GIs with hundreds of tanks parked against the hedgerows, along with artillery and amphibious landing craft. Every available space had been taken for stores and ammunition. It was extremely clever to keep such enormous amounts of equipment and supplies away from the coast, as the German spies (of which there were many) could not tell their masters where they were destined for, and yet they were blatantly visible. However, by this time the Allies had complete mastery of the skies and this equipment was not in any great danger from air attacks.

It became difficult to drive along the minor roads and lanes as the tanks, in particular, took up so much room. The tanks were very hard to steer with their caterpillar tracks. Added to which, the Americans had a happy-go-lucky attitude and left them in a pretty haphazard way. They didn't always make the gateways and entrances to the Milton Hall grounds and consequently, they were constantly having to carry out repairs to the boundary walls.

Despite the enormous impact that the GIs made, people liked them, as they were a very affable bunch, and the girls certainly appreciated them and their largesse, which was apparent wherever they went. Everyone was in a good frame of mind as the war was now generally going well on all fronts, such as North Africa, Sicily and in the Pacific.

Back at school for the summer term did not only mean serious study, but also cricket and sports. The weather was perfect and the builders had arrived to do the more urgent repairs. Stonemasons were busy replacing the shrapnel-pitted and cracked Bath stone blocks on the St Paul's wing of the main building.

I had a visitor, which was an unexpected pleasure. I was taken out to a super tea at the Red House in the centre of Bath – I don't believe that it exists any more. What a dreadful shame!

I had met Peggy Chisholm before in Almería. She was Geoff's friend and I thought that it was very kind of her to take me out to tea. Most of the other boarders were taken out regularly, but Geoff was so busy and far away that he only came once during my whole time at Prior Park. I am quite sure that if he had been able to visit me more often, he would have done so. Anyway, Peggy did it in style, and we went by taxi to the Red House and back. Apart from Peggy's visit, that term was uneventful.

I couldn't wait for the summer holidays and each day was wistfully crossed off my calendar until at last I was on my way to Felmersham. Geoff always tried to get away for a few days whenever I was on holiday with Floss and this was no exception. He arrived a couple of days after me. He was thoroughly enjoying his work in the RAF; it was very important and hush-hush. I was to learn after the war that he invented the fog disposal system and was also engaged in the early stages of guided missiles.

We had several visits from Lord Ventri, who came to see Floss as he had been a friend of her husband Vincent. He was a most interesting person and an absolute fanatic about airships. He was already busy formulating plans to begin building a new rigid airship immediately after the war. I was able to let him see dozens of photographs showing each stage of the building of the R101 airship at Cardington. These I found in a trunk in one of the attics.

Some of the photographs were in 3D and one was able to view them with a special wooden holder which had lenses and a slide, into which you were able to insert the photo/postcard and move it to and fro until it was in focus. I learnt that the R stood for an airship having a rigid construction; that is, made up from an aluminium frame, which took many individual gas-filled bags tucked in between the guiders behind the outer skin of the ship. Previous to this form of construction, they were just glorified balloons with a motorised cabin slung under the belly and held by slings which encircled the girth of the balloon.

The first successful rigid ship was the R34; then came the R38. I have, to this day, the middle section of one of the propellers of the R38 – beautifully made from several laminations of what appears to be mahogany. The two largest rigid airships, built in Britain, were the R101 and the R100; the latter was ultimately dismantled and the former, as you know, crashed over Beauvais in France.

After Geoff returned to his airbase, following our chat with Lord Ventri, I felt that perhaps I could now talk to Floss about the R101 crash. Up to now, she had not discussed anything to do with her husband Vincent, the crash and the state funeral. I suppose that as it had taken place but a short time earlier, the pain and sorrow was still with her and she was obviously still missing him terribly.

After a dinner in early September, we had all changed for dinner as usual – Floss was very strict and determined to keep up standards, in spite of the war. You just had to clean up and change into a fresh suit, collar and tie for dinner. Before the war, it would have been a black tie without exception! But now it was a change no matter into what; it was for the sheer principle. We were served dinner in the dining room by Kathy in her parlour maid uniform, complete with her starched headgear.

Dinner consisted of several courses, but none were substantial as Sarah, the cook, had very little to go on with and had to stretch our rations over the various courses, no matter how small. This particular evening, soon after Lord Ventri's visit, I could tell that dinner was going to be special, with a drop of port and brandy to follow! It was then that Floss began the search into her past.

★

She had married very young, 18 years old, in fact. She produced some old photos and there was no doubt about it, she had been very glamorous. He, too, was very dashing in his naval uniform. He had come from an upper-middle-class family and Floss had been a Hodder prior to her marriage – a well-known family in the printing and publishing business. When they married, Vincent was working for the air ministry, but went on to hold the naval rank of lieutenant commander, which was peculiar at the time. After many deliberations, the government decided to go ahead and build the most ambitious and largest airship of its time and Lieutenant Commander Richmond had the task of designing and building the ship at Cardington, where a second colossal shed was to be built to house it. The project went ahead, but typical of all governments, penny-pinching was the order of the day, and it would not allow for the more powerful and very importantly, lighter engines to be used as Richmond's original design had required.

Floss felt sure that this would have been a very important factor and probably the major contribution to the crash, as the ship did not have the power to clear the high ground during the storm. No doubt there were other considerations which would have had a bearing on the disaster as a whole. After listening to Floss for the best part of an hour, it became quite obvious that the politics and disagreements had played their part. It was the successful crossing of the North Atlantic by the airship R100 that put so much pressure on Vincent Richmond, but the R100 had been built with private money.

Floss continued to talk passionately and helped herself to yet another brandy. She and Vincent had been very much in love. He had had a very strong character, she told me, and showed me a photograph of him, pointing out his large bushy eyebrows. He had showered her with jewellery and anything that would beautify the house. He would regularly arrive home with a bouquet of flowers; he was a true romantic, she said to me. Her eyes were becoming moist as she tried to fight the tears. Her memories had been too real for comfort.

By any yardstick, the Richmonds had been comfortably well off. They employed a gardener, a chauffeur, a parlour maid, an under-cook and a cook, who was to all intents and purposes a housekeeper. They also employed a cleaner-cum-laundry woman. Floss had never raised a finger in the house, other than to see that everyone was functioning according to their established duties.

Due very largely to Vincent's type of work, they entertained on a regular basis, not only friends but also politicians, air ministry officials, senior management from Cardington and county people generally. From the government, Sir Seften Brancker, who was Director of Civil Aviation and Lord Thomson, Secretary of State for Air, were also frequent visitors and many other colleagues came too.

In a very sorrowful voice during her recollections, she said, 'We all knew that they should not have gone.' The weather forecast had been bad; a storm was gathering and Mrs Hinchlife, who was the wife of Captain W. Hinchlife – the pioneer transatlantic flyer – was predicting a disaster. Again there was more, but she suddenly said:

'Well, it really was down to two main reasons. Those heavy, slower and less powerful diesel engines; even though they worried Vincent greatly and he had tried to have them changed. Secondly, Lord Thomson insisted they go to India without further delay and didn't want to listen to reason.' We all knew that Lord Thomson was expecting to be the next Viceroy of India, which explained the latter.

Floss had seen the ship off, which included a galaxy of famous passengers such as Lord Thomson, Sir Seften Brancker and, of course, the crew. Flight Lieutenant Irwin, the captain, Wing Commander Colemore, Major Scott and Vincent all travelled as representatives of the Royal Air airship establishment and also as senior directors for the project. Major Johnston, whom Floss liked very much, was the navigator.

Floss said that she was informed of the disaster in the early hours of the morning by a telephone call from the air ministry. She was informed that only six people had survived and that, as far as they knew, her husband had been asleep at the time of the crash. She continued to talk with great feeling to me, and said that even before the news she had laid awake for hours because she had not liked the way the great airship had left its moorings. She had seen the pilot was finding the ship difficult to control, as the outer skin of it had already become saturated by the heavy rain storm, which of course increased the weight of the airship considerably. However, as they passed over London, a radio message had been received saying that all was well. This reassuring message did not make her relax and Floss had continued to worry through the night. It appeared that in the appalling weather, the airship had not risen sufficiently and because of its increased weight and underpowered engines, it did not have time and space to clear the hills at Beauvais.

With tears making rivulets down her cheeks, Floss recalled the moment she knew of Vincent's death.

'They were only able to identify Vincent by the wristwatch that I had given him. It appeared that he was wearing it at the time, but it was shattered and partly melted, as it was made of silver. But there was enough left of it to show his initials, VCR.'

Next Floss began turning the pages of some of the magazines that she had taken out of one of the study cupboards. They were filled with full-length pages of the state funeral, the gun carriages with the flags draped over the coffins. I could see in a couple of the photos that it was Floss with a black veil over her head and partly covering her face.

I could see that she was becoming more upset by the minute… the memories had been too vivid, so I decided that enough was enough.

'My word, look how late it is! You have a meeting at Carlton School tomorrow to decide what you are going to do with those boys who are being brought before the board of governors, so I think bed is beckoning, don't you?'

I went to bed.

Whether I had disturbed something or not, I don't know, but I now could hear footsteps coming up the rather creaky attic stairs. It was definitely spooky as they went past my bed and through the closed door to the room beyond, which was a boxroom full of cabin trunks and discarded furniture. I can't say that I was particularly frightened though; more intrigued I would say, as it was so real. I decided to investigate the matter further and consult the book that Martin had lent me on the history of Felmersham and its vicarage. I discovered that the house was supposed to have been haunted by a servant girl who had thrown herself over the banisters to her death from the landing just outside my room! It appeared that she had become pregnant and the son not only refused to take responsibility for her condition, but he told her to make an excuse about a serious illness in her family and leave at once. Because she realised that she meant nothing to the young master of the house and that her own family would have very little truck with her, she decided to end it all, we are told, by shouting out his name as she fell to the ground floor. In the still of the night, I often listened for the voice, but I never heard it. I only heard the footsteps, time and time again.

Chapter Ten

September 1943. The Allies had landed in Italy and all appeared to be going well. I was very chuffed, as I had just been made captain of the first rugby team. The season was to prove very successful as we won all our matches, except against Downside College, with whom we drew after a very hotly contested match where fortunes had changed many times. I was awarded my rugger cap, but because of the war I had to wait to have it presented to me when the war was over. Despite my several reminders and promises, Brother Burke failed to deliver. To be honest, I was rather miffed about it!

New Year 1944 and again Geoff succeeded in getting some leave. He was now squadron leader and we were all, of course, very proud of him as he was shooting ahead at the age of 51! We all expected the Allies to land in Europe very soon and we discussed my leaving Prior Park and finally agreed this should happen at the end of March 1944. I was to volunteer for the Fleet Air Arm; under the officers' training Y scheme, before I received my call-up papers.

I went for an interview at Cambridge. The scheme at the time was headed by Admiral Richards. The medical had shown that I was slightly short-sighted. This, I was assured, would not prove a problem as there were many pilots who had to wear contact lenses, and they appeared quite anxious to get me in.

The Allies had landed in Normandy and Cherbourg and we were all most anxious that they could get a sizeable bridgehead in order that the bulk of our forces should land safely. The first few days proved all-important, as they had to break out of the narrow beach-heads. This eventually, they succeeded in doing, but at a considerable cost.

★

At the same time, I went back to Prior Park for the last time. In many ways it was a sad time for me. I said my goodbyes to my housemaster, the head and, of course, to the many friends I would be leaving behind.

The journey home was quite uneventful. By now, we had become used to the flying bombs, or doodlebugs, as these things were generally known. They were pilotless flying objects, which had a very simple jet engine, and when the fuel ran out, they just dived to the ground. RAF pilots very quickly found that they could shoot down a fair proportion of these over the English Channel where they caused no harm.

Not long after the launch of the doodlebugs, Hitler released the most menacing V rockets. These travelled faster than the speed of sound. You were unable to hear them coming. The first thing you knew of them was the enormous explosion... too late to do anything about it.

The train I had caught from Bath to London was a couple of miles or so from Paddington Station when it came to a halt. A V rocket had exploded but a few minutes earlier at the entrance to the station. This had left an enormously deep hole, destroying four lines coming into Paddington. We were made to stay in our train for three hours whilst an assortment of railway personnel cleared a path for us to walk from the train, which had now pulled up to fifty yards or so from the edge of the crater made by the rocket.

By this time it was dark and eventually we were allowed off and were guided by railway wardens with hand-held torches. It was quite a caper to clamber down and cope with your luggage at the same time, without the help of a platform. It was a sizeable drop! Many passengers had to be helped off because of their age and fear of having to step out into the dark.

★

In October 1944, I returned to Emmanuel College, Cambridge and was boarded out in a terraced house facing Parker's Piece, on the east side of my college, which also faced the University Arms Hotel, where I had many a half-pint. Both were within easy walking distance.

I had a final medical and I was assured that the necessity to wear contact lenses would not bar me as a Fleet Air Arm pilot; several pilots were already wearing them. The officers' Y training scheme was a combined university and service arrangement. In my case I was based at Emmanuel, under Admiral Richards. I had to take a Latin exam at the Senate House and I followed this with the mechanical science tripos exam, known as Little Go. This was normally taken after the first year at university.

Everything appeared to be satisfactory and the senior tutor, Mr Melbourne, was happy to accept me after an interview with him and Floss. He was very impressed with her and became most interested when he learnt that I had been captain of rugby at Prior Park.

Victory in Europe was celebrated on 8th May 1945. Trainee pilots who had been most recently accepted were automatically discharged if not 100 per cent fit and this was a terrible shock and disappointment to me and to a great number of others, made worse when Geoff told me that the state bursary which I had gained was inadequate to cover the total cost for my stay at Cambridge. I was in due course to join him in Devon after his demob as he had bought a house somewhere in Exeter. I was gutted!

Floss was devastated too, at the thought of my having to live so far away and having to give up my place at Cambridge for which most youngsters would have given their ears. Floss offered to make a contribution, but it was all in vain. Geoff would have none of it. On reflection, I believe that money was not the only consideration; I think that it was Floss's considerable affection for me that made Geoff somewhat jealous and fearful that I would continue to live in Felmersham, and consequently he would lose me for good. Cambridge was far too close to Bedford for comfort! The tussle over Cambridge and my separation from Floss was extremely hurtful to me at the time. I suppose that I never really forgave Geoff for it, but on the other hand, many wonderful things that followed my move to Devon would never have been possible if it hadn't been for this decision. If Floss had had her way, I would have stayed at Cambridge and I am sure that I would have got my degree. Probably, I would have gained my 'blue' in rugby too, but none of these things would have compensated what I would have lost by not coming to Exeter. The funny thing is that, if I am honest, I still perhaps resent my abrupt departure from Cambridge just a bit.

I was now back at Felmersham, but not for long. The refugees had returned home to London. For some reason or another, unknown to me, we moved quite suddenly to the Old Rectory at Pavenham. I knew that Floss had only leased the house at Felmersham and it may be that the church commissioners wanted the house back as the actual vicarage again. And in fact, that is what happened to the house at Felmersham; within two or three weeks of our exit, the vicar and his family had moved in.

Pavenham Rectory was very much bigger and built on a grand scale. It was constructed from hand-cut limestone blocks, and was probably not quite as old as Felmersham Vicarage. It was very spacious – it had a coach house with accommodation over it and it was one of those rooms that I took possession of immediately, making it my den and workshop. The house was blessed with beautiful large stone mullioned windows. It was in every way an impressive edifice, with an equally important stone arched entrance that had massive double oak-panelled and studded doors.

The main house had 11 bedrooms, of which 4 were for servants as they were not so finely decorated as the others. In addition to these bedrooms, there were 2 over the coach house, but the toilet was downstairs at the rear of the garden. Strangely enough for a house of these proportions, you would have expected to find several toilets, but in fact there were only two, one at each end of the main corridor upstairs and one for the master bedroom leading off from the adjacent small dressing room. The servants had to cope with large decorative chamber pots under their beds.

The house had drive-in-and-out iron gates and a sizeable shrubbery in between; not quite like Odell Manor, but impressive enough to suit Floss, as appearances were paramount to her. The garden was not as large as you would imagine, but it was big nevertheless and had a hard tennis court with a suitable high fence surrounding it.

The house throughout had very deep dark oak skirting boards

and the study/library was oak-panelled from floor to ceiling. The fireplaces were made from veined grey marble and they were large. It had a central heating system installed at some time or another, as there were pipes and enormous cumbersome radiators everywhere but the boiler, which was supposed to work on solid fuel, was never lit in my time there. Fires were lit in the study and sitting room most days and occasionally we had a fire in the dining room, especially if we had a dinner party. Everyone had hot water bottles in their beds and in winter, I kept my socks on, as the bottles never warmed the bottom of the bed. Sometimes I had to make a trip to the loo in the night, and walk along the linoleum floors, which was so cold as there was no wall-to-wall carpeting in those days!

We had more visits from the Liddell twin brothers, as they now lived only a couple of hundred yards away. Nellie, Mabel and Temperance continued to live with Floss. I think that Temperance's time in the Women's Land Army helped her to come out of her shell; she would now join in conversations and actually have a good laugh when the penny dropped. She did in fact have quite a dry sense of humour. Russell stopped coming to see her as, at last, he got the message.

I began to feel that Floss's finances weren't too healthy. She had been poorly advised on some stock market decisions and was let down badly by Argentinian Railways. She was trying to put a brave face on it, but she definitely suffered a crippling financial blow from which she really and truly never recovered – too many eggs in one basket.

In one way and another, she felt very vulnerable and hence never missed an opportunity to pressurise me into staying with her, as she considered that she needed me to take care of things. She continued to try to talk Geoff out of my going to Devon and said that it was still not too late for me to return to Cambridge. She had recently had words with Mr Melbourne and he was able to confirm to her that that was the case.

With money problems looming, Floss tried another tack and came up with a new plan. I was to work for the marine engine manufacturers; no less that Allens of Bedford, where she had an influential friend on the board. She was more than confident that

I would get a senior PR job with them. She was so confident that, after a cursory interview, I was persuaded to take a course on mechanical engineering and mechanical drawing at Bedford College of Technology.

Little did Floss realise that her attempts at keeping me in Bedford at any cost by encouraging me to go to the technical college, would have a dramatic effect on my life. I became totally infatuated with engineering and physics. I had discovered that I had an engineering bent and developed an insatiable appetite for anything pertaining to engineering design. From now on I spent a considerable amount of time trying to invent things that would be useful and for which there was a need.

I started several prototypes, of which some proved useless, but some others went on to make money! Friends often brought me things to repair as they either didn't trust other people or thought that they could have it done on the cheap. Anyway, word got around, no doubt helped on by Floss, to the effect that I could manage to repair most things. But if not, I could probably give some useful advice! One has to remember that, because of the war, there were very few people available to do the odd job or two who had not been called up, and also, you could not get spare parts either, so you had to improvise.

Geoff came to stay for a short leave, prior to being demobbed even though the air ministry had offered to keep him on because of the very important research work that he had been engaged in for the RAF, now that he had the rank of acting wing commander. The first part of his leave he spent visiting Bristol, Exeter and Bideford to see about his old job when he came out of the RAF. He told us that Hillsborough had been occupied by the Americans as an officers' mess and that huge areas of Northam and Westward Ho! had been put out of bounds. Apparently hundreds of landing craft had been stored there as well as it being used for exercises on the Burrows at Westward Ho! and Braunton, before the Allies landed at Cherbourg and Normandy.

Geoff was eventually demobbed in June 1945 and we had a general election on 26th July 1945. It brought a new government – Labour – for the first time, and with a huge majority. Clement Attlee became prime minister. I was quite sad to see Winston

Churchill resign; he had been such a wonderful wartime leader... What a shame. Geoff went off for an interview for the post of chief engineer, South Western Electricity Board. All the electricity companies, no matter how large or small, had been nationalised by the new Labour government. The whole of the south-west, from Cornwall to Gloucester, and of course Devon, were now included under the umbrella of the south-west area board, of which Exeter was a sub-centre. Geoff took over responsibility for Cornwall and Devon and a large slice of Somerset. The executive appointments at area level were mainly political and were appointed according to the government of the day.

On 6th August 1945, America dropped the first atom bomb ever, on Hiroshima, Japan. The whole of the naval city was virtually wiped off the face of the earth. On 9th August, just three days later, the second atom bomb was dropped, on Nagasaki. The devastation was immense and the loss of life was in the tens of thousands. These terrible bombs brought war with Japan to an abrupt end. The Allies accepted Japan's surrender on 14th August 1945. We had two national holidays so that everyone could celebrate. I tried to forget about the suffering that we had inflicted on the Japanese people, but we also had to remember what they did to our prisoners and it did also stop the war, which otherwise would have gone on maybe for years. The Japanese didn't give up easily and to surrender and live another day was looked on as cowardice in their culture.

For some reason or another, we did not have the same sense of relief and pleasure that we all felt on VE Day. It had to be that horrible feeling of guilt at the mass killing of innocent people – women, children and all. It lurked uncomfortably in everyone's minds. But on the other hand, no more fighting and killing in Burma, the Pacific and Malaysia. Surely that alone had to balance out our terrible deed?

*

It was a pleasure for Geoff to join his old friends, particularly Will Smith, whom he had worked with for so long before. Both had to resign from Whitehall Securities Corporation in London. Will

Smith had already been appointed chief accountant for the same sub-area as Geoff. I was pleased at this too, as I had known his family so well from the Almería days.

In September 1945, Geoff bought a small semi-detached house – No. 2 Salmonpool Lane, Exeter. As the name implies, it was in a road leading to the River Exe. Geoff wrote and told me that I should join him in Devon as early in October as possible. I was torn in two directions; I was looking forward to being with Geoff again, but I had also become very attached to Floss, and she to me. So whatever I did, I was going to hurt someone in the process. With mixed sorrow and misgivings, I left Floss and Pavenham. I felt a twinge, too, for Temperance, as we had become very good friends. I had just done an oil painting of her. We were like brother and sister to each other and I am sure that we both knew that our attachment would never flourish into a full-blown relationship, as our feelings for each other would not prove strong enough to withstand a long period of enforced separation.

I knew that Floss would be devastated when the time came to say goodbye, and she was. She pleaded with me to change my mind, but eventually she too realised that I had to make my own way in life. It was very much a man's world then and I had to face it on my own, without her and the comfort of a Cambridge degree under my belt, which could have been mine so easily.

By the time that I got to St Pancras and then on to Paddington Station via the underground, I had begun to look forward to seeing Geoff again, the new house and of course, a whole new world and an entirely new life. The dice had been cast.

Chapter Eleven

It was several years since I had last been in Exeter, and here we were just coming into St David's Station. The platform was jammed with people, however, I could still see a beaming face, full of expectant joy. The last time I had seen Geoff look like that was when I arrived at Gibraltar in the summer of 1936. We went to the car park where he proudly showed me a brand new car, a Standard 12, black with beige interior, very smart. Well, it was new. You had to be in an essential or very important job to qualify for a new car so soon after the war. There was very little civilian traffic as petrol was also severely rationed. We received several envious glances as we went through the town.

No. 2 Salmonpool Lane was a bit of a disappointment, as I had never lived in a semi-detached before. In fact, I had never lived in a row of houses... A culture shock, you might say! Anyway, it would appear that it was a very desirable house as it had been built by a very reputable builder, none other than Stavertons of Dartington and because of this Geoff had to pay over the odds for the house... £985!

The initial shock was considerable, but the inside was something else. How on God's earth could two grown-ups ever live in this tiny space? You could have put the whole house in the sitting room of Odell Manor! Geoff had obviously not had time to get much furniture, which was just as well, as there was very little room left for it when the two of us were in the front room. The dining room had already been turned into a workshop.

It appeared that you could only buy new furniture on coupons and priority was given to newly married couples, so we did not qualify on that count. A guardian and a ward I am sure were never on this list of qualifications. I learned later that this furniture had a kite mark; not exactly luxury, so you were better off to buy second-hand stuff though the overall effect was somewhat depressing – especially to someone with a background like mine.

The people on either side of us were unobtrusive, as the English always used to be. In fact, they may as well have not been there at all for the number of times that we saw them. But their presence could nevertheless still be felt. We had never lived so close to anyone before and it took a bit of getting used to, believe me. I had the feeling of being watched all the time. As we shut the door you could generally detect a slight movement of the curtains of one house or another and it was all a bit weird. On the left of our house, we had Mr and Mrs Rae whom we rarely met within the first couple of weeks, as they kept very much to themselves. On our right, in a much superior house to ours, lived Mr and Mrs Haile who occasionally managed the odd grimace and perfunctory 'Hello', but both would duck and weave or turn away in time in order to avoid eye contact. And so we found out very little about our neighbours for the first six months. The Hailes had a pretty sickly son, who did try to be sociable whenever our paths met, not by speaking but by giving me a side look and a knowing nod. He obviously knew a thing or two, but he was not going to let on what that might be! Maybe he uttered a dozen words during the whole time that we knew him and his family. We could always tell when he was home as we could clearly hear the xylophone, which he played in the front room. Actually he was quite good at it, if you like that sort of thing.

I suppose that we presented a weird picture to most people. They found out after a time that Geoff was a bachelor and therefore, I was obviously not his son unless…? As time went on, we were able to discern the odd smile or two. Yes, they had fathomed that I was Geoff's ward and any previous sinful assumptions were forgiven.

Geoff was well aware of my initial disappointment with the house and tried to make things work out for us. He really put a lot of effort into it. But he did not have to have worried one little bit, as Salmonpool Lane was to prove not only a milestone in my life, but also as the foundation to a happy and contended future which fulfilled more than any aspirations that I might have had. I was lucky and I count my blessings every day, without fail.

Well, first things first. I had to get myself a job… Ugh! Not a proper job – Will Smith had persuaded Geoff that I should be an accountant as, after all, I was supposed to be bright – at least Geoff thought so. Horror of horrors... I couldn't think of anything worse. My interests lay in designing and making things. I knew that I had a flair for thinking out fresh ideas, to the point of taking out patents for my inventions.

My protestations were unheeded and I became an articled clerk to the senior partner of Ware Ward & Co., chartered accountants in Cathedral Yard, Exeter. I was gutted to learn that Geoff had to pay for my articles. An occupation that I detested and, to make matters worse, I was to work for the next three years without pay! I could foresee a life of utter destitution and I was not at all happy with my lot, not by a long shot!

Early in October, I went to the county ground and met Mr Combes, the secretary of the Exeter Rugby Club. I also met Mr Roebuck, the team secretary. Because I had been captain of rugby at school and I also had my colours, it was agreed that I should come down to the county ground and practice on Tuesday and Thursday evenings after work, each week. I was selected to play for the Harlequins two weeks later, as right centre in the threes.

Ralph Ware was a kindly sort of person who was very much the product of the public school system. We got on very well. His wife was very different, an absolute tartar. She would burst into his sedate office unannounced and demand money with menaces. We could hear her telling him that she was broke and didn't know where the next cocktail party dress was coming from. Mind you, his daughter was no better either and flaunted her good looks at every opportunity, as well as caning her father with equal ferocity and menacing ways as her mother. Poor old Ralph! He didn't really have a dog's chance.

Other partners, such as Bill Curtis, for example, were on the whole okay. They tolerated us but looked upon us as a bunch of semi-idle useless morons who were unfortunately articled to them or to one of their fellow partners.

All the articled clerks were boys; girls had not as yet arrived on the scene. The only women in the office were typists, secretaries or account clerks. None of them were capable of turning your head unless some of the boys became a bit desperate. I had to start a postal study course to cover the theory part of the job. Much as I had expected, I found the subject dreary and uninteresting, but kept plugging away for Geoff's sake. But it was really like flogging a dead horse.

On Sundays, we would more often than not go to the Seven Gables in Beech Avenue, Pennsylvania. And as Mrs Smith was such a good cook, we looked forward to getting a super high tea. Each time we sat down to our long anticipated feast, Will Smith would spoil it by asking how I was getting on with my accountancy studies. Mr Smith did his best to encourage me by pointing out how well accountancy had served him. Well, he did certainly appear to be pretty well off.

The mainstay at Ware Wards was the chief clerk, old Reg Hurford, who was not particularly enamoured with the likes of articled clerks, as they were far too independent for his liking. However, as the saying goes, his bark was definitely worse than his bite. In fact, he could be quite amenable at times. Ralph Ware, on the other hand, was a sort of figurehead that only dealt with the top clients after some underling had done the donkey work.

The real driving force of the partnership was Bill Curtis, who made it his business to search out prospective clients at Exeter Golf Club and the Exeter and County Club, better known by the name of The Musgrave, of which I too became a member, many years later. Bill was also a director of Tinley's Café, in Cathedral Close. For this reason, we had to ensure that he did not see us having our coffee every morning! We failed to avoid him one morning; he was not too pleased to see us, as he did not expect his privileged underlings to be sitting in his café drinking their coffee on the firm's time, instead of drinking the dishwater served up as coffee in the office.

Although accountancy was not by any stretch of the imagination an enjoyable activity, I did like most of my time at Ware Wards. They were a friendly bunch and the atmosphere was great. I enjoyed meeting clients, especially farmers. They had vivid imaginations and would avoid using books wherever possible, which was a large part of the time. Bartering was the generally accepted system.

Clients who handled cash were in a class of their own; they had creative accountancy minds. They would look astonished, even hurt at any suggestion of having pocketed the money without making an appropriate entry in their haphazard and somewhat casual system, which they called book-keeping. I noticed that all purchases were religiously entered, plus a few more for luck! This, of course, made it particularly easy as I worked out the notional gross profit by using a well-established set of ratios for each type of trade.

I often had the feeling that they thought I was only working for the Inland Revenue at their expense. Despite that, they were most generous and I would often find gifts left on my desk, especially from the farmers when they learnt that I lived alone with Geoff and we had no woman to look after us.

The office parties at Christmas were an excuse for everyone to let their hair down, and occasionally more so than they had intended! Many secretly repressed feelings were released in the popular filing room at this time of year, even though the room was full of rather sharp metal cabinets. It wasn't the most comfortable snogging equipment around, but then the lights often failed and that made up for a lot!

After the Christmas holidays, the most senior members of staff endeavoured to forget their outrageous and uncontrolled flights of passion – several current romances came to an abrupt end and new ones took their place.

In my early days in the firm, I had a mild flirtation. It really was a truly platonic affair. She was very attractive and was well aware of it. She found my lacklustre and wishy-washy attention to her inadequate, so it was not long before a much more willing and amorous suitor came onto the scene. Not only was he handsome and debonair, but he was also a willing victim to be sacrificed at the altar before the following spate of Christmas parties. I am glad she got her man as, sadly, she died a few years later.

Work generally was routine and boring. In order to relieve the monotony, we would use elastic bands and fire paper clips from our first-floor windows in Cathedral Close and hit unsuspecting people on the park benches, who were either enjoying a peaceful rest or eating their packed lunches. On many occasions, the target would be Mahatma Gandhi. We used to see him regularly between 1945 and 1946, riding his bicycle.

Gandhi would rest his cycle against the back of a bench right in front of our window, so he was a perfect target. I must say he took it very well, so much so that we got to like him and we all became nodding acquaintances. We had no idea that he was a religious person and that he would one day became world famous during his struggle to get India free. He was, in fact, a Hindu lawyer from a middle-class family.

He wore a loincloth and used to sit cross-legged next to his bicycle in contemplation. As far as we knew, he wore nothing under his loincloth. However, no matter how hard we watched when he got onto his bicycle, we could never tell!

The only exception to the monotonous work tasks we had, were the ministry jobs. These were interesting jobs, and for the impecunious articled clerks, generous out-of-pocket expenses were a gift from heaven. Our itemised expense chits were so infrequently checked that we became more adventurous and imaginative as time went on. The ministry jobs consisted of visits to dairies and brick companies, etc., in order to ascertain the cost of a pint of milk or a brick, and so on. These cost exercises were carried out by various accountancy firms up and down the country at the same time. The results were compared by the particular ministry concerned, adjustments made for any peculiar or special circumstances and a unit price for say, the pint of milk or the brick was established. Basically, everything was controlled and where necessary, subsidised. Only the black market was uncontrolled. Anything in short supply and not on coupons was generally available if you knew the right contact and were prepared to pay the going rate. The rationing system, on the other hand, was good, as you were sure of getting what you were entitled to at prices fixed by the Government.

Ware Wards was quite a big firm by West Country standards, as it had several branches dotted about the whole of the South-west. The staff was made up from an incredible mixture of people of varying backgrounds; especially the articled clerks. They came not only from the UK, but also from the Commonwealth and colonies. One chap for example, was larger than life; always cheerful and kept a notebook, which contained every joke that he had ever heard. He was fun to work with and I learnt a lot from him. He was very clever and always appeared to have time on his hands, as he finished his work well within the allotted time, while the rest of us were nearly always being chivvied to finish a particular job.

There was another character whose life apart from accountancy revolved around anything fishy... He wrote about it, gave talks about it and also made the most exquisite flies. He was just a born fisherman. He certainly did not bother too much with his appearance – his shirts were invariably covered in the most colourful handmade flies. You certainly avoided any comment about any particular fly, for you would get chapter and verse about where and when to use it!

Then, of course, there was Old P... What a case. He was the product of a well-known public school. He too kept a book – a betting book. We all placed our bets with him on a regular basis; sometimes we even had an accumulator. Because of our enforced poverty our bets were necessarily rather small, but we did nevertheless bet on most races each day. Old P sometimes did not lay off our bets with Dollar Beer with whom he was privileged to have an account. No – he decided on many occasions to take on the risk himself, especially if he didn't think much of our choice!

As we placed bets on most races, at the end of the day we were very rarely much up or down, so Old P made quite a modest return on us. It was quite tragic to see Old P allowing the old enemy get the better of him, especially those chasers – little baby bottles of Red Label Barley wine. To be fair to him, he had a very bad time in Burma with the 14th Army. He was sent to Trincomalee for a rest after a gruelling spell in the jungle. Whenever I saw him looking at the wall in front of him for any length of time, he would smile and say that he was suffering from another attack of 'Trinco Stare'. I thought that some of his yarns about his experiences in the jungle were a bit far-fetched. However, I was to find out that they were, in fact, accounts of real encounters. One day, we had a new clerk join us; he too had been in Burma, in the jungle and what's more, he and Old P recognised each other as old buddies from the war in the Far East. They had not seen each other since they had been demobbed.

I will call our new colleague T. He had been badly shot up in a ghastly encounter in the jungle and had only recently left the hospital, where he had undergone massive face-rebuilding surgery. It was good, but it was still difficult to look him straight in the eye, despite the wonderful job that the plastic surgeon had done... His face was not a pretty sight.

It became quite obvious that P and T were close friends and it turned out that P had been T's sergeant. One night in the Burmese jungle, T had been left on duty as P had unfortunately led his small reconnaissance party into an advanced Japanese patrol. In the middle of the night, 'P' woke up to the sound of a gun. Crack!... Crack!... Crack!... And again... Crack! Although furious, immediately fearful for the lives of everyone, he said in a stern but low voice, 'This will mean a court martial. Make no mistake.'

'Yes,' came the reply from T, with a very resigned sigh.

P cautiously led his small section through the undergrowth and made a quick recce. Much to his surprise and amazement, he found four Japanese bodies.

'Well, you have saved our lives... But it will still mean a court martial. Do you understand?' said Sergeant P. Fortunately that threat did not materialise and a medal for bravery was the ultimate outcome.

The Japanese were past masters at jungle warfare and could move without disturbing a single leaf. After so many years, the 14th Army also learnt not only to master the jungle, but to use it to their advantage. T's hearing must have been so acute that he had been able to kill those Japanese soldiers without seeing them and without wasting a shot.

I got to like T a lot; he was such a warm person, not the sort to go around killing people, but I regret to say that his nerves were just the same as P's and this was eventually to pull them both down into the abyss of a civilised cesspit of emotions.

I am still in contact with D, the fourth member of our group. I met him again recently, after many years, at a Masonic function and reminisced over our happy times with Old P and T. He had a good brain, is good company, a *bon vivant* and altogether a good *hombre*.

★

How lucky I was to have had the good fortune of meeting and working with such grand fellows. Several of us became interested in tennis and went regularly to Heavitree Park. None of us were exactly a Fred Perry, but we enjoyed our get-togethers. Some were from my circle of rugby friends and several came from the office, which also included some of the girls. What a transformation... White shorts, very short. Somehow they were so different on the courts that it was hard to believe that they were the same girls as at the office!

However nice these girls might be, at that precise moment I was rather preoccupied with the daughter of our next-door neighbour. I had only just caught a glimpse of her, but what I saw was rather good, to put it mildly! A bit of a cracker was she! Must make urgent contact, that was definitely my top priority. My wishes were to come true, sooner than I expected and what's more, it was to be in the most favourable of circumstances and not after a tennis match, as I had first hoped. It was after a beer session with the boys, following a rugby game at the county ground.

We had all ended up at the Elephant and Castle in North Street. Of course, it doesn't exist anymore. In fact, it was pulled down to make room for Sainsbury's and Woolworths. I remember the night as if it was only yesterday, even after five or six pints of best bitter, for I was just about to put down my half-eaten pork pie when I noticed that the inside was covered by a thick layer of green mould. I started to shout at the barman, holding up my pie in order to complain, when the police rushed in, telling us to stay as we were! They asked if any of us had been up stairs in the last hour or so. Nearly all of us had been to the gents toilets at the back, which were situated at the bottom of the stairs. We were all questioned individually, when there was a scuffle and a young chap whom I had never seen before was handcuffed and taken away.

We had heard a scream above the normal hubbub of the bar, but no one took the slightest notice. The rumour around the bar after some of the police had left with their prisoner was that a very pretty girl, whom most of us knew, had been stabbed. Her family owned a newsagents and tobacconist business in Queen Street.

Shortly after an ambulance arrived, but they appeared not to be in a hurry. When the stretcher went through the saloon bar, it became obvious that she was dead, as the body was covered from head to foot. She had been murdered!

The whole place became hushed and instinctively we spoke in undertone. The more you thought about it, the more incredible it became... She was being murdered as we stood there drinking without a care in the world, except for my rotten pie. It was quite a shock. As far as we knew, she was engaged to the landlord's son and often served in the saloon bar. The police made the landlord close up and we all went to the pub opposite.

At the trial, the young man was sent down for her murder. I believe that she had been a bit free and easy with the lads in the bar, but no more than the usual *piropos*, as it is called in Spain, such as 'Hello gorgeous!' No harm intended, just letting the lass know that her looks were much appreciated! But, obviously, someone had become very jealous.

I can't remember why I should have ended up in Countess Weir. Anyway, I decided that it would be better for everybody if I was to drive my old Ford Pop along the grass verge in Topsham Road. Unfortunately, it had several drain ditches across my path right up to Salmonpool Lane, by which time I was rather shaken and certainly the worse for wear.

The car went up the concrete slope in front of the garage. Some how it stopped, or should I say stalled, just before hitting the double doors. I was distinctly unwell... I lay back on the seat, I listened. I looked around for bedroom lights to come on. Nothing. Thank goodness.

Perhaps I should sleep it off in the car? I could see the lights from the street lamps next to me, but it was utter darkness on the passenger side. Oh dear. I could just make out the small privet leaves pressed against the glass of the passenger door window. 'Oh my God. Things are not looking too good,' I said to myself. But what on earth was I to do?

There was no way that I could leave the car in such a position until morning, but it already was the early hours of the morning! It was quarter past two. With only one thing to do, I backed the car gently down the slope and reparked it further away from next door's hedge. Yes that was it... But wait a minute. I knew that I had definitely hit something. No, on second thoughts, perhaps I had imagined it. As I started to reverse, it became quite clear that I was catching something that kept going 'twang'. It went twang several times...

Oh dear. The beautiful privet hedge lay on the lawn... and we hadn't even met its owners yet! I had no idea what these new people were like, but I had a feeling I was soon going to find out quite soon.

Sobering up at a prodigious rate, I decided to wake them up and offer my apologies at once. I pressed the bell and it gave a most alarming sound at that time of the night. I stood in front of the door in a state of utter panic. A light went on in a front bedroom immediately over my head. It seemed an age before I heard the lock catch being pulled back.

'Oh hello, Mr Northcott. It is Mr Northcott, isn't it?' I said, sheepishly.

'Yes, what do you want at this time of the night... Is there something wrong?' he answered.

I explained that I had unfortunately ripped up his privet hedge, and he asked me what on earth I had done that for, to which I did emphasise that it had been an accident. The rain was now easing off, but everything was soaking wet. Without saying another word, Mr Northcott went over to the felled part of the hedge and proceeded to heel into the soft ground, each privet root, one by one, with his carpet slippers on! I joined in enthusiastically.

In no time at all the hedge appeared in one piece again and looking no worse for its experience; in fact, it looked perfect. He stood looking at the hedge and then turned to me.

'I don't think that it would be wise for your to be seen in your present state, by your...' said Mr Northcott as Mrs Northcott appeared at the front door. I believe that she had seen all the goings-on from an upstairs window.

'What you need is a strong cup of coffee,' she said, with a distinctly friendly smile on her face.

They turned out to be lovely people and weren't a bit angry with me. They just took everything in their stride. Well, I had certainly made an impression on our next-door neighbours... Not exactly a good start! A large cup of black coffee was carefully directed into my outstretched hands. I sat on the edge of one of their sitting-room chairs, looking rather sheepish as I could feel their eyes fixed on me, probably wondering whether it had been such a good idea to buy a house next to such odd people. Nervously, as I put down my cup on the flat wooden arm of the chair, a gorgeous, shapely figure of a girl came into the room. She had grey-blue eyes, fair hair and walked tall in a closely wrapped dressing gown in pale blue satin fabric with large, deeper blue flowers. Luckily, I had put down my half-empty cup as she entered the room.

'Oh, there you are,' said Mrs Northcott. 'This is our daughter, May.'

I got up and shook this beautiful girl's hand.

'My first name is Helena,' she said. It could not have been anything else for such a gorgeous creature. Rather feebly I said 'Mine's Don. I am delighted to have met you. Sorry it had to be this way!'

I explained what had happened. My old Ford Pop had sharp springy front bumpers, the ends of which stuck out, actually quite dangerously. As I had gone up the ramp, the end of the bumper on the passenger side sprung past each privet root and pulled it out of the ground. I finished my coffee and again apologised profusely, looking at each person in turn before making a quick exit.

As I opened our front door, I could hear Geoff in a deep sleep... Thank God. He had been oblivious to it all. I was very relieved as he could be very stern at times and certainly this would have been one of them.

★

Sunday morning, not feeling too bright. It wasn't surprising really, after my escapades of the night before! My dinner jacket didn't look too good... I had a vague recollection of my first annual rugby dinner, but I did remember who had been there; there was the president, Dr Ashford, plus Jack Combes, the secretary, and Roebuck, the team secretary. Our special guest was the mayor of Exeter, Mr Tarr, who obliged us with a spirited dance on the top of the Top Table without falling off, even though he was by that point, three sheets to the wind! Our other guest was John Maud, MP for Exeter. Paul, our captain was also there, of course.

It was quite a nice morning that morning, especially for March. I was slowly coming round and the sun was actually shining. I thought to myself that I must keep an eye open for Helena. Geoff was in his workshop / ex-dining room. He seemed to be okay.

It was about this time that Geoff had his aforementioned change of name. I have no idea why I should have called him Charlie Boy, but he appeared to like his new nickname and gave me one of his special hugs.

'What's up then?' he said. 'Had a good night, *mijito*?' That was his term of endearment for me; it was Spanish for 'my little one'. Well! I explained that yes, I had had a good night, and had also managed to meet the people next door, especially the daughter, called Helena, who was quite a cracker! Charlie (Geoff) managed to emphasise at this point that I shouldn't forget that I had my studies for the intermediate exam a few weeks from now.

He carried on brazing some small component for the 3½" gauge railway engine that he was making, whilst we were talking. It was fortunate that he did not ask me when I had met them, or for that matter, when I had come in. Apparently he had not been feeling too good and had decided to go to bed early and read one of his very intellectual books. He had dropped off to sleep long before 10 o'clock, thank goodness. I did not have to tell him any little white lies. I wasn't good at it and it always made me go red in the face.

Feeling a great deal better, I swivelled my armchair around to face the bay window, so I could see anyone coming out of No.1. I sat there for a considerable amount of time and I was beginning to wonder what to do next, when I saw a Pekinese dog through the garden gate. Attached to his lead followed the gorgeous girl from next door. Helena! I was in luck... Just a quick exit to the main road.

'Hello. Going for a walk? Can I come with you? I don't suppose you know the walks by the river at the bottom of the road?' I said.

'Well, I do, as a matter of fact' she answered, 'but you can show me more if you would like.'

I found out that her dog's name was Pepper, but I don't know how much of the conversation she could hear, as Pepper never stopped yapping the whole time.

We didn't have an exhilarating conversation. In fact, it was all a bit of a rush. Apparently she and her parents were going out to tea at some farm or other. Unfortunately, I did have enough time to learn that she was already engaged to a captain in the paratroops. Not a very good start, I thought. However, I could sense that my interest in her and enquiries for future contact were not being discouraged out of hand. In fact, I had the feeling that if I played my cards right, I could be in with a chance!

I found Mrs Northcott very pleasant. She invited Charlie and me to a high tea very shortly after. She was a superb cook. Within a week or so, we were both invited again, this time to a fantastic Sunday roast. I have never eaten so much in all my life. Charlie gave me a couple of funny looks; I knew I was letting him down, but hell, we were having a pretty rough time with the standard of cooking at our house. Well, that is if one could call it cooking, which more often than not caused an attack of foul fiend indigestion!

Mr Northcott was also pleasant. He loved his garden, but regularly incurred the wrath of Mrs Northcott because he simply loved to do his gardening without wearing a shirt. I could tell that Helena was also none too pleased with his stomach being exhibited to all and sundry, but she was too fond of her dad to let it become an issue.

Charlie had organised a twenty-first birthday party for 5th June 1947 at the Manor Hotel at Moretonhampstead. The Smith family came to my party and I would have like to have asked Helena to come too, but I felt that she had not known me long enough and might have thought that I was being too forward – especially as I had to remind myself that she had been engaged for quite a long time. Charlie was not very forthcoming either, as he was anxious to avoid female distractions so close to my impending exams.

It was a great pity that I didn't ask Helena to come as I would have enjoyed it so much more, but there it was. I suppose that just at that moment I didn't want to rock the boat, but on reflection I should have had the courage to insist on her joining us. I was mortified to learn from her, some time later, that she had been expecting to be asked and was quite hurt when she hadn't been. It turned out to be our very first tiff!

I was becoming more infatuated with Helena by the day and tried to see her on a regular basis, but if that failed I had to resort to throwing pebbles at her bedroom window in order to make contact, which fortunately faced onto our back garden.

Shortly after my twenty-first, I went to London to sit for my chartered accountants intermediate examination at Church House. Unfortunately, I failed two out of the five papers, but unlike today, I would have to take all five papers together again in six months. Charlie was naturally disappointed as he thought that I had the brains and ability to pass, but what he did not realise was that my heart wasn't in it. I had no zest for accountancy and in any case, my mind was for the moment preoccupied with other things such as Helena, whom I saw more and more.

I was now able to see Helena at will, for she had got herself a fantastic new job. She was now the area manager of King George's Fund for Sailors. It was more like fun for sailors! She had a very posh office in Barnfield Crescent, right next to our family solicitors. It was very handy for me, as I was able to walk over from Ware Wards in Cathedral Close at lunchtime and evenings, or for that matter, any time!

I found myself being drawn to Helena whenever I had a moment or two, but rugby, oddly enough, was still my number one preoccupation, and practice at the county ground on a Tuesday and Thursday evening was sacrosanct. On Saturdays, I was either playing at home or away. This of course was in winter and spring... It was just as well that I didn't play cricket!

One night, I took Helena out for a drink at the Royal Clarence but went into the cocktail bar, which wasn't the 'in' place at the time. One had to wear a tie which was a bit of a bind, but well worth it, as it was never crowded and Helena could sport all her gear. The popular short drink then was a gin and orange, and if you were lucky you got a lump of ice as well. You always did at the Royal Clarence.

We lost all sense of time and wandered home by the most devious route imaginable, stopping now and then to recharge our batteries. We finally made it home some time after midnight, and were shocked to see all the lights on in No. 1 as we turned into Salmonpool Lane. Helena used her key in case her parents had left the lights on for her on purpose. As she pushed the door open she found herself staring at a tall figure in military uniform standing stiffly to attention. The cane under the arm was the only thing that was missing!

We both had a look of horror on our faces.

'I didn't know you were coming,' said Helena, as cool as a cucumber. I did not need an introduction… Helena's paratrooper captain looked exactly as I had imagined… LARGE! He extended a hand in order to introduce himself, which at first I mistook to be the first dose of medicine that I was about to receive from him. But instead, he was as nice as pie. We went into the hall, where Mrs Northcott stood with a distinct smile on her face. I got the feeling that she was pleased that Helena's fiancé John had caught us together and so late at night.

I didn't stay long as I was somewhat speechless and could only manage a sort of grin. It appeared that John had been given some unexpected leave and proposed staying for a while. However he must have assessed the situation to be unfavourable and left the next evening. He continued to visit Helena for quite a while, but as time went on, it became patently obvious to him that things were not going to be the same between him and Helena, added to which, Mrs Northcott didn't like him particularly. She found him rather overpowering and boisterous, and encouraged Helena to break their engagement. You would have imagined that John might have become embittered, but on the contrary, we became the best of friends instead.

My visits to Barnfield Crescent increased in frequency. Often I would be sitting working in the office when I would have a sudden urge to have a nice cuddle, so off I trundled to see Helena in the hope that she would be in. Lovely cuddles and whatsits followed by sorting out the petty cash book which had got into a fester. Probably the stamp book also required straightening out, and then of course, were the collection boxes to be emptied and the tens of thousands of copper coins to be counted. I felt obliged to help as I used to take up so much of her time, generously given for my benefit on other things!

Mrs Northcott I liked very much. She was about to receive one of my nicknames, as has always been my want. She was given the tag of 'Dickey' or just 'Dick' for short. She was a good sport and quite liked her new and unusual name. Very much later she had a further addition to her name and for good reason.

She became very trusting and allowed me to visit Helena in her bedroom when she was suffering from asthma and bronchitis. After several visits, I found it better to get into bed with Helena in order to have our cuddles and whatsits. It was rather chilly and after all, No. 1 Salmonpool Lane had no central heating. I didn't mind catching anything from Helena. It proved a good move. Mind you, it did get a bit crowded when Pepper insisted on lying at the foot of the bed. I encouraged Helena to stay away from work until she was fully recovered… I had my reasons!

Chapter Twelve

Charlie took delivery of a Jaguar 1½ litre in racing green. He must have ordered it many months previously, but had kept it a secret. It was an absolute beauty. It had a radiator very much like a Bentley. It had a walnut dashboard, rev counter, green leather upholstery and wire-spoked wheels... Wow! We were the envy of everyone who saw it. Undoubtedly, it must have cost a *paquete* as they say on the Costa del Sol! This particular model was to prove the success story of Lyons, the chairman of British Leyland. Naturally I was anxious to drive it, but Charlie couldn't bear to part with it until the novelty had worn off.

Eventually I was allowed to drive the new Jaguar, at a fortuitous time as it happened. Helena was giving me the runaround with so much attention from her Royal Navy friends and the adjutant of the Royal Marine camp at Lympstone, Captain Benton. He would turn up in full dress uniform to impress her. On top of that, he would show up in his chauffeur-driven staff car, just to show me up... My transport consisted of the old and battered Ford Pop.

As luck would have it, that particular morning, Charlie had very reluctantly agreed to let me drive his spanking new car. Poor old Benton. We arrived simultaneously at Barnfield Crescent. I couldn't believe my eyes... There he was riding a bicycle in full dress uniform, complete with spurs!

Never saw poor old Michael again... It was too humiliating for him. In fact, sadly he died very suddenly a few years later. As far as I was concerned, he was number two written off; however, my problems were by no means over. I still had some formidable hurdles to circumnavigate. Helena had been in the American army and many of her officer friends, whom she assured me had been and were absolute gentlemen, kept on keeping in touch and I found this most irritating.

I suppose that Lieutenant Commander Burnett was my most immediate concern; an adversary that was going to take a lot of guile and skill to overcome, for he had very considerable resources as well as being well connected. Yes, undoubtedly he was a most dangerous individual. His uncle was none other than the current Commander-in-Chief of the Home Fleet; on top of which, his family was titled and had estates in Scotland. To make matters worse, he was also captain of the HMS *Devonshire*.

The main trouble at the time was that Helena was such a gorgeous-looking girl that all the male talent she came across was without exception, attracted to her like a powerful magnet. It had become a full-time job for me to fight off the competition, so much so that I had to adopt a cunning and cavalier attitude of careless abandon. I obviously did a good job, as Helena believed it to such an extent that she became totally mystified... Especially when she saw me taking out a couple of girls. I felt like a complete skunk, but it was after all, in a good cause, all said and done!

Now and again, Helena had to go to Admiralty House, Plymouth, in order to report to Admiral Burnett who at that time was chairman of King George's Fund for Sailors. Often I would offer to take her there and to various other meetings, as I was always able to slip out of the office without undue trouble.

It was one of her meetings at Admiralty House that I offered to drive her to Plymouth. I sat in the car near the front door to the house, as I was not suitably dressed, to say the least! I was listening to the radio when the equerry to the admiral came over to my car, saluted very smartly and said, 'Are you Mr Donald López?' For a moment I thought I was going to be under arrest! 'The Admiral and Lady Burnett would be delighted if you would join them for afternoon tea,' he said.

Well, you could have knocked me down with a feather... This was so embarrassing. I looked like a navvy; they had probably assumed that as Helena looked her elegant self, I would also be adequately dressed. I was suitably announced and I received a very warm welcome... Perhaps I had noticed a slight quiver of Lady Burnett's eyelids, but both were most cordial. For a growing rugby lad, the tea that was served could only be described as a ritual, and was not meant to satisfy anyone's hunger pangs, certainly not mine! Each of the assorted sandwiches measured 1½" by 1½" – just a mouthful. I did not attempt to see through them, but they were pretty thin! I managed a couple of rich tea biscuits, before declining the offer of a further sandwich or another small wedge of delicious fruitcake… What a fool!

The admiral was now on his feet. He asked me to accompany him and started to talk about the different oil paintings which were hung around the room. He singled out one in particular. It was an oil painting of the HMS *Belfast* with its forward gun turret firing a broadside at the German pocket battleship, *Sharnhost*. The admiral stood up to his full height, all 5' 8", and said in a proud voice, that he had been the commander of the *Belfast* at the time when the German warship was sunk.

He was about to relate the account of this very important battle at sea, when the equerry came into the room and gave the admiral a smart salute before saying that the HMS *Vanguard* was about to pass between Admiralty House and Drakes Island. The admiral appeared to be quite flushed and, to my surprise, asked me to follow him into the garden. He climbed the rostrum at the end of the garden and then turned and asked me to climb up a step and stand just behind him so that I could have a better view.

As the *Vanguard* approached, we heard the bosun's whistle and the admiral immediately took the salute. Two seconds later we heard the sound of the first gun, followed by 16 more. The admiral looked quite short, and I could see from where I was standing that he had obviously enjoyed his pink gins by the number of visible chins. He held his salute whilst counting each gunshot aloud.

HMS *Vanguard* passed by with her officers and ratings lined up, standing to attention, which was no mean achievement as the strong undercurrents just at this point made the huge ship list to starboard. Salutes were cut and I could see that the admiral was fuming as he turned to me. I had no idea what on earth could have upset him so much!

'I'll have his guts for garters! As Commander-in-Chief of the Home Fleet, I am entitled to a nineteen-gun salute!' he snorted.

He had hardly finished speaking when all who had been standing to attention were back on deck with what appeared to be

153

a rather whimsical smile on their faces. Suddenly there was a loud boom, rapidly followed by another boom.

'I should bloody well think so,' muttered the admiral under his breath. He then apologised for his French, which he explained just comes out, whenever he is under any sort of bother. I reassured him jokingly that my French wasn't very good, as we stepped down from the rostrum. We said our goodbyes to Lady Burnett and made our way back to Exeter by the longest possible route, which gave Helena ample opportunity to thank me personally for taking her to Admiralty House.

★

Because of Helena's job, she was in constant contact with the Royal Navy in one form or another, but especially with the other Burnett. She was invited to cocktails, etc., on HMS *Devonshire* and of course he was not only there, he was the bloody captain! I had no alternative but to swallow my pride and I drove Helena down to Plymouth again. I had to wait for her until the party was over, but I did feel that this was beyond the call of duty. Fortunately, this time he overplayed his hand and I was soon able to eliminate him from my short list of adversaries!

However, it was not long before another beggar appeared on the horizon. This time it was none other than Lieutenant Commander Karrens, the national hero of the River Yangtze incident! Now this fellow was not going to be a pushover. He, showing tremendous skill and courage, took HMS *Amethyst* a couple of hundred miles or so up the River Yangtze, in the middle of communist China, at a time when diplomatic relations did not exist. This trip was in order to rescue some British and other nationals. It was bad enough on the way up the river, but coming back, the Chinese were more than ready and he had to run the gauntlet, passing the heavy guns from either side and smaller but murderous gunfire from each bank, all the way to Shanghai and eventually to the open sea.

As you can see, this chap was not only courageous and brilliant, but he was handsome and also a national hero at a time when Great Britain needed him the most. We were at low ebb at this stage of the war.

Lieutenant Commander Karrens had the looks of a hero and most certainly enjoyed his triumphant return to the old battered country, where he was fêted from Land's End to John O'Groats. Helena was always on the lookout for a popular celebrity and he became an obvious must for her. Helena had very persuasive powers, but Karrens was easy meat for her as he was to act as the honeypot for one of her fundraising balls at the exclusive Fortfield Hotel, Sidmouth.

The lieutenant commander was scheduled to speak about his dashing escapade up and down the River Yangtze, but his mere presence proved enough to make it a sell-out. As usual, I attended the gala event, but on this occasion I was an invited guest, so I had to sport a dinner jacket. As I did not possess such a decadent item, I had to buy one from very meagre resources... Nevertheless, of good quality.

Helena very kindly allowed me to invite a couple of friends, so I asked David and Henry. We were to be as good as gold and helped to sell tickets for the grand draw. His nibs was to officiate and, apart from presenting the prizes, he would also give the girls a peck on the cheek. Some preferred a full kiss on the lips, as it turned out!

When the lieutenant arrived, you would have imagined that he was royalty. The place was in turmoil in no time at all. He was even better looking than we had all thought and he had a wicked smile, which seemed to captivate even the grey-hair brigade. I could see now how he had been able to carry out his audacious exploit up the Yangtze without a paddle... He was born for the part!

No sooner than he was able to break loose from his adoring throng of girls than he was off chasing Helena along the upstairs corridors and bedrooms of the Fortfield. Helena put up a genuinely good fight of it, which I think took him by surprise this time. Nevertheless, Helena's call for help sounded desperately urgent, so I was off like an eager beaver and soon found the bedroom in which he had cornered his prey. My looks must have been enough, obviously more than the Chinese had been, as he stopped trying to smother her and said, 'Okay, you win.'

Despite my apparent easy victory, I felt that he should be told a few home truths and reminded him that he was not at this precise moment on the HMS *Amethyst*, and that in any case, Helena was a much more difficult proposition to cope with than bloody Chairman Mao! I told him to behave from now on or my stalwart companions and I would forcibly throw him out, which I said in a very stern voice. Unfortunately, the most my stalwart companions could do was to attempt to stay upright on their feet by that point. Luckily, Karrens took the warning on the chin and from then on he behaved in an exemplary manner for the rest of the evening.

As the number of would-be suitors never appeared to diminish, I had no alternative but to ask Helena to marry me, but for now she would have to just consider herself engaged to me... The ring would have to come along later, when the finances were up to it. As it turned out, however, the ring became an immediate must. In spite of my financially depressed state, we bought a most expensive engagement ring by any normal standards; for an impoverished student, it was practically a mortal charge on any future earnings. The bank, fortunately, proved most obliging, or should I say, the manager became a close friend. I am not too sure whether it was a genuine friendship, or whether he just wanted to keep a close eye on his investment.

The new ring certainly put the brakes on Madam, and as such, I felt that any thoughts of marriage had to be put on the back-burner for the foreseeable future. The trouble with Helena, if I can put it that way, was that she had had a most exciting life in the American army. It had in fact, been more than exciting, it had also been most enjoyable!

During the largest part of the war, Helena had been a private secretary to a three-star American general, based at Wilton, Salisbury, which at that time was being used as a German prisoner-of-war camp. She was thoroughly spoilt by the American officers who were mainly from the wealthy landowner brigade and who knew how to treat a lady when they saw one. Helena was certainly one of these. At the time she was at Wilton, it was the same time that the Americans were still treating their Negroes as sub-humans; their pets were better off. The Negroes were made to do all the menial tasks, serving white officers at the table, and in fact Helena had a Negro standing behind her chair, attending to her every need.

With such a background, you must be wondering why a gorgeous girl who could have had anything she wanted should have become infatuated with a foreigner like me! Well, I don't know either, but I thank my lucky stars and tell her every day that she is a perfect angel. Even our early years together must have tested her love and tenacity of purpose, particularly when we were so strapped for cash. Perhaps she knew, as I always did, that one day I would make enough to live comfortably in our later years.

I took my intermediate exam again and, as luck would have it, I passed the two papers that I had previously failed with flying colours, unbelievably, only to find that I had failed one of the subjects that I had passed the first time round. I was very much aware that Helena was a considerable distraction, but I don't believe that it was the sole reason for my failure. I lacked the will power and interest in the subject. To put it mildly, my effort was so lacking in lustre from start to finish that there could have been no other outcome. I disliked accountancy as a profession as I found myself acting as a servant to those who could afford my services, due to their success in life. I wanted to be on the other side of the table. Apart from that, the work itself I found to be very boring, as nine-tenths of the time it was just checking the accuracy of hundreds of figures, lots of adding up and finally getting to the bottom line showing how much profit the client had made for the Inland Revenue to tax him on. These figures were about other people's achievements. This was not for me. I wanted to create and lead. I decided that enough was enough. I predicted that at some time in the not-too-distant future, I would have my own business. My accountancy experience was, however, to prove a very useful tool. It meant that from now on I would be able to control my own future and heed the warning signs when things were not going to plan.

I had to get away from Ware Wards and get experience in a large organisation. Luckily a position came up at the sub-centre of the South-west Electricity Board. I got the job despite the fact that Charlie was the sub-area chief engineer. The union, NALGO, was none too pleased with my appointment. Fortunately, as time went on, they realised that Charlie had had nothing to do with the appointment, and for that matter, that I had very little contact with him at work. I was eventually made a local NALGO representative for the Exeter centre.

I proved my worth and quickly had further promotion into an area entirely away from Geoff's domain. My new job was to locate all the equipment, stores, transports, transformers and switchgear, etc., at the precise takeover date. I had to record the whereabouts of this vast quantity of equipment, which had belonged to numerous private companies, the day they were taken into public ownership, i.e. nationalised. You have to remember that at that time there were no computers to help with the vast number of items. The only help we had was a punch card system which consisted of semi-stiff buff cards with holes punched into them by a machine much like a typewriter. The position of the holes gave the description and part numbers of each item. It was a time-consuming exercise to sort into similar part numbers as you had to pass these cards through a sorting machine which could only deal with one digit at a time.

Work had obviously continued after the war. In fact there was a big spurt forward to make up for the lack of development during the war years. New electricity lines were being constructed in every part of the country at the same time. We had the money to do so as the USA were lending us money on very favourable terms under the Marshall Plan.

Electricity and generation was divided into various area authorities. It was a gigantic task. Each of the numerous private companies had different recording and accounting methods. They had a mixture of old and written down plant as well as brand new and very expensive equipment. I had to evaluate everything that covered the whole of Devon, Cornwall, parts of Somerset and Dorset. My superiors were more than pleased with the result. I was within a couple of hundred thousand pounds when all was said and done. This was peanuts when compared with the whole capital involved at nationalisation of the South-western area.

Helena and I had now been engaged for about three years and Floss was not at all happy about it. She did her best to make me go back to Bedfordshire, especially after she learnt that I had given up accountancy. She made a point of coming down to Devon and took the opportunity of introducing herself to Helena. Neither of them took to each other. Floss was in one of her worst moods and called Helena a mere office girl. This naturally made Helena smart and angry. Floss told Helena that she was certainly not good enough to marry me! I am afraid that this time Floss had gone too far and we had a few sharp words.

She returned to Bedfordshire the next day. We did keep loosely in touch by the occasional letter, but I was never to see her again.

★

Helena was becoming rather impatient and was putting a great deal of pressure on me for us to get married, especially when she found out that the matron at the Princess Orthopaedic Hospital had invited me and the rugger friends, which included Henry and David, to the monthly dances at the hospital held for the benefit of the young nurses. Frankly, we three went for the free beer and nothing else. Unfortunately, the matron got wise to it and our super free evening came to an end after we failed to dance with the nurses following the final warning from the matron. She made it clear; no dancing, no free beer. So that was that!

Charlie and I bought No. 1 St Leonard's Place, just up the road from Salmonpool Lane. It was a beautiful house; Georgian, detached and very large, but in a bad state of repair. No. 2 Salmonpool Lane sold very quickly and Charlie and I moved to St Leonard's immediately, despite the mess that it was in. This move caused a certain amount of aggro, as I was now unable to see Helena on a regular basis. Helena's parents gave us a very expensive bedroom suite in anticipation of our pending marriage, so it was delivered to our new house at St Leonard's, where we had ample room, to say the least. Charlie and I had bought very little furniture to date, as the Salmonpool Lane house was very small.

Unfortunately, our relationship became stormier by the day; so much so that we broke off our engagement and Helena had the bedroom suite collected by Mark Rowes and stored by them for an indefinite period. This move caused considerable resentment, as the house now looked as if no one was living there and I felt that my lovely girl was being very spiteful.

Fortunately, things did not stay in such a limbo stage and before long we were both glad to kiss and make up. In no time at all, the bedroom furniture arrived, without notice as if by magic! I got to know the removal men, whom I felt were somewhat sympathetic with my plight. On reflection, I feel that we should have done the right thing and invited them to our eventual wedding... This would have made them more settled in their minds about us!

★

I had no idea what on earth had hit me. The wedding day had arrived.

Helena had, with her organising skills, engineered the actual day with me noticing nothing. After all, we had been engaged for about four years, so I wasn't exactly on the lookout for any untoward signs. Charlie had become very concerned for me as I am sure that I looked quite alarmed. He kept repeating, 'Do you realise you are about to get married, in just a few hours? You had better get yourself together.' He said this as he put a large, neat whisky into my wet, floppy hands. 'There you are, you will need this if you are going to see the day out...' More or less his final words to me before the curtain fell!

Frankly, the fact is that I do not remember too much of the actual ceremony. My recollections are very vague, except that I remember Helena coming up to me and standing by my side and I thought how beautiful she looked, before I technically passed out of the real world. I remember nothing more until we got into the car and on to the Imperial Hotel for the reception. Now this I remember well, as I thought I was the luckiest person in the world... And of course, it turned out to be so.

I have a feeling that officially, I never agreed to marry Helena; certainly not in the normally accepted sense. I gather that I only managed to make a squeak when asked if I would have her, to have and to hold, etc. Despite my very poor showing as a groom, Helena made up for all my shortcomings by looking stunning, radiant and beautiful.

At the reception, everyone thoroughly enjoyed themselves and I actually joined in the jollifications as well, sort of swept away by the occasion and of course, the sheer relief that it was all over, bar the shouting. Helena's uncle, Dr Cecil Northcott, made a rather intellectual speech – I suppose that he felt it was expected of him in view of the fact that he was a Cambridge don as well as being general secretary of the Congregational Church and a writer of travel and children's books. He was also the *Daily Telegraph* religious correspondent and a regular speaker for the BBC on religious matters.

Helena's aunt, Mrs Jessie Northcott, was at one time a tutor at Girton College at the same time as her husband, Dr Northcott, was a lecturer at either Newnham or Fitzwilliam College, Cambridge – I can't remember which.

The bridesmaids were absolutely lovely. The eldest, Joan who was an old school friend of Helena's, emanated from Jersey and was great fun to be with. The other bridesmaids were the Mansfield sisters who lived opposite us in Salmonpool Lane and both were very pretty. The youngest was quite a case. Whenever I think of them, I remember the little one, as she would do handstands and show her knickers. She would often do this, in fact, most times whenever I went by. I had that sort of effect on her, it seemed! Their mother and father were charming people. The father had an antique business in North Street, Exeter, where he also had a thriving undertaking business, which he combined with a great sense of humour… All very odd!

Although I had an apparent unpreparedness for the serious business of marriage, which Helena had organised to perfection, I did a good job on the honeymoon arrangements, utilising every available penny at the bank to ensure that we had an exciting and hectic 10 days in London over the Christmas and New Year festivities, with a theatre show pre-paid for every evening.

We had the usual chaotic send off from the reception at the Imperial. Helena wore a stunning tailor-made two-piece bottle green suit with a pencil skirt… She looked absolutely gorgeous. We caught the evening train from St David's station to Paddington, first class, and were lucky enough to find an empty compartment.

The ticket inspector gave us a knowing look and suggested that we pull down the blinds in order to warn people and hopefully put them off getting into our compartment. This didn't stop him checking on us a couple more times. I think that he was rather hoping to catch us out!

Dinner on the train was absolutely fabulous as we received so much attention from the dining car staff and I am sure that we had far larger helpings than normally served; we did feel very happy and no doubt it showed. We had a lot of knowing smiles from people. We were delighted with our double room at the Cumberland Hotel in Oxford Street. When we entered the room, we found a large bowl of flowers. Attached to one of the flowers was a tag saying 'Congratulations', signed by Madame Marcelle. She was a very well-known French dressmaker and had made Helena's wedding gown and going-away trousseau.

It was to be practically the only time in our lives that we slept in a double bed. I think that the excitement and sheer tiredness had been too much for Helena, and her carefully worked-out monthly calculations went adrift. But it wasn't a worry. I was, in the words of a psalmist, absolutely knackered!

We both slept like newborn babies. It was wonderful to find Helena curled up close to me when I woke up in the morning. It was Sunday, 23rd December 1951. We went down to breakfast and tried to look casual, like an old married couple. We were obviously rumbled very quickly, but it was a relief to find that we were not the only honeymoon couple in the dining room. Some looked terribly washed out; on the other hand, we were in fine shape and ready to do London!

In order to ensure that our time in London was free from any financial worries, as I said earlier, all theatre shows had been paid for in advance for most evenings, when we would return to the Cumberland with just enough time to wash and change for dinner. On a few occasions, we stayed at the hotel to enjoy the dinner dances, especially during the Christmas period and New Year's night. For these evenings, it was dinner jackets and evening gowns only. Much to our delight, on some evenings we would find flowers or champagne or both on our table from well-wishers, who would catch our eye and raise their glasses to drink to our future happiness, and sometimes these well-wishers were a considerable distance from our table. I believe that on New Year's Eve there were over 1500 sitting for dinner! The dance floor was in constant use and as at this stage I had not had dancing lessons, I was very glad to get lost in the crowd, although Helena was a good dancer already.

We had packed so much into our holiday that we were quite worn out and could have done with a rest when we returned home, instead of having to go to work immediately. It was a bit of an anti-climax. However, we needed the money as I had emptied the coffers for the honeymoon, although funds in London lasted out pretty well. We did, however, still have to skip lunch on occasions and fit in another show, or have lunch and return to the hotel early and stay in during the evening. This was no hardship though, as there was always something going on.

Unfortunately, Granny Websdale had ensconced herself at No. 1 St Leonard's Place whilst we were away on our honeymoon, so Helena came back to Exeter not only to look after and cook for me, but also Charlie and Granny Websdale. It really wasn't fair, but Helena never complained. She was just simply wonderful and had boundless energy.

It was only some years ago, when Charlie died, that she did say that Granny Websdale had been very thoughtless in staying with us from the first day of our joint lives together, and if that had not been enough, she would invariably behave in the most aggravating manner. She never offered to help and expected to be waited on hand and foot. She had another irritating habit of whispering to Charlie right in front of us. He would look most embarrassed. She was highly critical of everything that Helena did; never a kind word crossed her lips and she was always on the lookout in case Helena had missed something.

'You mark my words,' she would say, 'a new broom always sweeps clean... Nothing new about that... But it won't last long, you see!' Oh, what a ghastly person she was.

Weeks went by and I pleaded with Charlie to get his mother to return to Southsea. I will give him his due; he did try many times, but she would appear to be stone deaf to his suggestions.

Chapter Thirteen

In order to augment my meagre income, I had the idea of turning one of the large rooms in the basement over to growing mushrooms. They had now become very popular and were also expensive to buy, 10 shillings per pound. As it happened, this venture did not prove frightfully successful, but I did grow enough for the odd sale or two to the local greengrocer and quite a nice lot for the family. We did not offer Granny Websdale any mushrooms as we thought that they would prove too indigestible for her. Well, she was so furious at not being offered any that she demanded to have a great helping in order to make up for the times that she had been missed out.

In the middle of the night I heard a great commotion. Charlie called me to his mother's room, saying that she was dying. As I entered her bedroom, a great spurt of blood came gushing out of Granny Websdale's mouth. It was awful.

We rang for Dr McKie at once and asked him to hurry as we didn't think that she would last long enough for him to see her alive. She kept haemorrhaging and we both took it for granted that the end was very near. Dr McKie arrived incredibly quickly and the ambulance only a couple of minutes after him. He must have arranged for it before he left from home.

Granny Websdale was carried out of her bedroom on a stretcher and as she passed me in the hallway, she raised herself on one elbow on which she propped her head and said, 'You won't get rid of me that easily... I shall be back. You wait and see!' Fortunately, she did not come back for a very long time, and then it was only for a few days. It was a great relief to us all, including Charlie, as it had been eight months of utter hell with her there.

*

We soon made up for the lack of passion opportunities whilst in London and we simply loved our new home in St Leonard's

Road, in spite of Granny Websdale and her constant carping. It was a most friendly house, huge by anyone's standards. There were five bedrooms; some were very large and, in the basement, there were servants' quarters complete with their own enormous sitting room. It was a semi-basement as the base of the windows were at the same level as the garden, hence the rooms were light and fresh. Originally, we had a butler's pantry, but this, we converted into a big playroom when our twins were born some years later. There was a blackboard, which took up most of one wall of the playroom, and under the large sash window, I built a full-length unit with open shelves to take all their toys.

Our bedroom had two large sash windows overlooking the City Basin and the Haldon Hills beyond; each of the windows had box seats underneath them. Helena was by now looking pretty large, as we were expecting our first child. Her size was causing Dick, Helena's mother, a bit of concern and she had considerable doubts as to whether her daughter was going to stay the full course of nine months from the date of our wedding. Her anxiety grew by the week, and although she was looking forward to having her first grandchild, she felt that the longer it took, the better! I am sure that she was not the only person who was studiously counting the months since 22nd December 1951.

Unbelievably, Granny Websdale decided that she would come and stay with us for a couple of weeks again. Whether Charlie forgot to tell her that Helena was six months pregnant, I don't know, but I doubt it would have made any difference if he had told her. She still expected to be waited on, in spite of Helena's condition. I would say that if anything, she imposed on us even more than before. The two weeks went by and she was still with us. At the end of five weeks, I gave Charlie an ultimatum. This time he was very firm with her and she went back to Southsea in high-dudgeon. She did not keep in touch with Charlie directly; instead she got her friend to tell Charlie that, in her opinion, his mother was not at all well and felt that he should come and see her as soon as possible. The whole thing was a put-up job, but Charlie wasn't prepared to take a chance and promptly went to Southsea, only to find that it was a false alarm.

You can imagine what a great relief it was to everyone, when the date for the minimum pregnancy period passed by... From now on, it would be sheer anticipated pleasure, especially for Dick. She could now look forward to the happy event. However, the first week of October went by and Helena was now actually overdue! What a change. Helena, on the other hand, was looking fabulous and worked herself to a standstill.

Well, the night arrived at last and Helena started to have labour pains but, being stoical as usual, she did not panic and waited until the pains became virtually continuous. Suddenly there was a great gush of water. Now I was in a real panic and rushed around like a blue-arsed fly. Helena has always been methodical to the nth degree, so as you would expect, she was already packed and I drove to Mowbray Nursing Home like a scalded cat. As far as I was concerned, the baby was definitely on its way and our old Daimler seemed to know the way, as I do not remember one little bit of the journey!

When we arrived, Matron met us at the entrance and chivvied everyone, once she knew that Helena's waters had broken. It was a bit of an anti-climax for me, as I was told to go home and ring in a couple of hours' time. I had a large drink and waited. I gave it two and a half hours just to make sure.

'I'm sorry Mr López, your wife is still in labour and there is no sign of the baby yet. Ring in the morning,' said Matron in a very calm and matter-of-fact voice.

I rang several times a day and each time got the same answer. Your wife is still in labour. It was unbelievable. By the fourth day, I was a nervous wreck. What about my poor wife? It was now the fifth day and I had not been allowed to go and see Helena. The poor girl must have been worn out, and we were all beginning to worry about her surviving, especially Dick who knew far more about it. She could not understand why she did not have a caesarean operation. 'You wouldn't treat an animal this way,' she kept saying. Dick kept on making the point that it was going to be a dry birth anyway, and that was going to be difficult enough as it was.

Miracles do happen now and again... I telephoned on the morning of the fifth day and was told to ring again in a couple of hours as things looked more hopeful... EUREKA! 'Mr López, you are the proud father of a baby daughter, weighing in at 9lbs 12oz. Your wife and daughter are both fine, but after putting up a great fight.' I was first in the ward that night.

★

Well, I couldn't believe my eyes. Helena looked positively radiant and our new baby girl, in a cot next to her, was the most gorgeous-looking baby girl in the whole world. I don't think that I have ever been forgiven for not bringing flowers to the nursing home like everybody else did… I just rushed out of the house and never gave it a thought! I hope that I have made up for it since; heaven knows, I do try! We definitely had to have the best pram in the world, it had to be none other than a Silver Cross, with huge high wheels, especially the front ones. They were gigantic. It was in a racing-green colour, with a matching gabardine hood and was beautifully padded inside, fit for royalty!

Naturally Helena was very proud of her new baby. We all were, but of course, she was the one that had to bear all the pain and suffering, and if for that reason alone, Rosalyn was extra special. We all thought that it was going to be a boy as Helena had become large so quickly and it was kicking all the time, and originally a boy had been our preferred wish for a baby. The only names that we had thought of were for boys. Helena had set her mind on Nigel.

You can imagine that it was quite a surprise to get a girl. Any thoughts of Nigel went out of the window, as if he had never been thought of. We now had a beautiful baby girl. We scratched our heads for a suitable name. I had recently been playing rugby against Rosslyn Park; this name rather influenced us, so with a bit of a change, it became the beautiful name of Rosalyn! Since then, this name has become very popular.

Our sitting room at St Leonard's faced south and it had two French windows opening onto a veranda, which ran the whole length of the house. At the centre, it had a flight of stone steps down to the front lawn. Rosalyn was regularly parked on the veranda after being fed, lying in her new and very posh Silver Cross pram for all to see as they passed by.

Helena simply loved work and had boundless energy. I do not know to this day how she managed to do so much! It used to wear me out just watching her. She would cook breakfast, prepare a light lunch for Charlie at midday and on top of that, she produced a three-course dinner for all of us each night of the week. If that had not been enough, she would also cook a load of cakes and pies and other such delights for the weekends, when her mother and father came to visit. On top of all of that, she would work until the early hours of the morning, redecorating the house, as it was in a pretty awful state of repair. The sitting-room fireplace, for example, had been painted over many times. Helena removed numerous coats of paint with the use of several strong coats of Nitromors to reveal a magnificent white marble fireplace, which had a most attractive blue tinge.

Not long after, Helena fell pregnant again and she had no option but to get help in the house, especially as in the meantime she had insisted on having foreign students from the International School nearby in 'the village' (as we called the area around Magdalene Road) in order to make a bit of money. We were at that time hard-up and could not see a light at the end of the tunnel.

Our first help was called Daphne, rather a quiet girl of 16 years of age. She was inclined to be rather morose, but she did work very hard. She had motherly instincts and became very fond of Rosalyn, as well as Helena's make-up and cosmetics, which she borrowed and kept in the bedroom under her bed. Helena would use them and put them back under Daphne's bed for fear of upsetting her!

★

It was quite pleasant working at the electricity board in Paul Street, Exeter. Fortunately, I had several rapid promotions and the money situation improved to the point that Helena did not always have to borrow money from Daphne no sooner than she received her wages!

Despite all the promotions that I had received, it became very necessary to find a job that paid considerably more money in order to make ends meet, having chosen to have a lifestyle well above the average for my age. Fortunately, my departmental boss, who was also the local organiser for the Exeter tennis tournaments and talent spotter, had a friend who owned a large luggage factory, making the then well-known Kingfisher brand. This friend had asked him to keep an eye open for a bright young man who could become his personal assistant.

My boss's friend was looking for someone who could take over the everyday running of his business whilst he was otherwise tied up at the city council, as he was at that time sheriff of the city. The company was called Gilchrist and Fisher Ltd. It employed about 300 in the factory and 22 in the office. It was still, however, a family business. He was the managing director, his father was the chairman and his mother did the designing as well as being a director and the true driving force behind the business!

My chance came when the general manager decided to emigrate to Southern Rhodesia and my boss at the electricity board considered that I would be the ideal person to combine the job of general manager and personal assistant to the managing director, Councillor WH Bishop, CBE. I have a feeling too, that my boss had been unable to keep me 100 per cent busy and, consequently, I was often up to some mischief, causing him considerable embarrassment, particularly as he did not wish to reduce his establishment and loss of seniority as a result. He arranged a meeting, which unfortunately I was unable to keep, as I had to go to the hospital suddenly, at Poltimore, with kidney problems.

Some weeks later, I was delighted to learn that Bill Bishop had been willing to wait until I had recovered. It appeared that Norman Weaver had done such a thorough sales spiel on my ability and general character that no one else would do!

I took to Bill right away and I was hopeful of getting the job that he had described to me at our first meeting. It was just the sort of job that I would relish. Luckily, my background and manner suited him and I could tell that he too had taken a shine to me.

Suddenly he said, 'When can you start? The job is yours and the starting salary is £600 p.a.' I was somewhat startled and took a while to answer. Had I heard him correctly? I was only getting

£385 p.a. at the time, which was average for my age and experience, but this would be nearly double! Whilst I was hesitating, he said again 'All right, damn it! We'll make it £750 p.a.' I am positive that it was my hesitation that made him think that I had not been happy with his first offer. Without further ado, I got up and shook his hand before he changed his mind... My hand had become rather hot and sticky. I said in a higher-pitched voice than usual, 'I am delighted to accept the job and I am sure that I will enjoy working for you.'

I told Mr Bishop that I had to give a month's notice but that as my boss was a close friend of his, he might be able to let me work out a shorter notice. As it happened, the board was about to reduce the administration staff as, by now, the initial takeover and amalgamation traumas were behind us. Things had settled into a routine operation, so the management were quite happy to accept a two-week notice and I started my new job in October 1953.

I was over the moon; I had never had so much money. I really felt rich. We bought a car, a Daimler, second-hand of course. It had a pre-selector gear-change mechanism – quite a collector's piece. It was a lovely car to look at and Helena took to it like a duck to water.

As the BBC was still using our coach house for their BBC2 booster station, we had to build double garages in the back garden to house Charlie's Jaguar and our recently acquired Daimler. My new job didn't really require the use of a car, but I felt that Helena needed to be mobile. In fact, my new job was only a short walk down to the River Exe, where the old Kingfisher factory was situated. I loved it and wished that we could have stopped there for good. My office faced the river and I had a large elm between my window and the river. Many a time, I saw kingfishers perched on the branch close to the factory window; in fact, Bill's father took a fantastic photo of a family of kingfishers, all on a single branch. This we used for many years as our advertising material.

In January 1954, we moved to our new factory in Western Way. To celebrate the opening of the new factory, we invited the more important customers to a super factory party. Wine and champers were on tap the whole day. This was before the drink and drive restriction had come into effect, which was just as well!

Because of Bill's council connections, we had the Mayor, our MP, a crowd of city councillors and of course, our nearby and more important customers, who came from all parts of the country as the company was very well known for its Kingfisher luggage. Bill was the leader of the council at the time, as well as being sheriff, for at that time, Exeter was not only a cathedral city, but also a county in its own right. It therefore had a separate police force with a distinctive uniform.

The new factory was easier to operate than the old premises by the River Exe. I liked working for Bill, and we got on with each other from the start. Although Bill initially continued to be the managing director of Gilchrist & Fisher Ltd, I soon become more than just a personal assistant. In fact, his council work and his love of entertainment took up a great deal of his time, so that within six months or so, I was literally running the whole show single-handed.

St Leonard's proved to be more than just a house to us; it was a real home with lots of character and a sense of grandeur. We even had our own ghost, built in, so to speak! We discovered that our ghost was friendly and loved the house as much as we did. We never found out whether it was a he or she, but whatever we did to the house, s/he appeared to approve and obviously could not bear to leave the place. The very first time that we came across our friend the ghost was one night when I woke up and felt something heavy on top of my feet. I took it for granted that it was old Fluffy, our gorgeous, loveable and totally adored tabby cat. He was more like a dog than a cat. Just at that moment, Helena woke up as well and sensing that I was awake, said, 'There is somebody in the room, can you feel it?'

'No,' I said, 'but something is pressing down on my feet, and it doesn't feel like Fluffy.'

'Yes' she said, 'I can feel an atmosphere… It woke me up.'

I switched on the bedside light and saw the dressing-table mirror directly in front of us, between the two sash windows, shake quite distinctly, as if someone had walked across the bedroom floor. The strange thing was that I could now move both my feet!

The ghost problem, if that is what you might call it, gave us quite a few disturbed nights. It was not unusual for the two of us to wake and hear someone moving about downstairs. At first, I was convinced that we had a burglar and naturally, each time that this happened I would creep down the stairs without making a sound. But each time it turned out to be a false alarm. However, more often than not I would find the same pair of French window shutters open, and half of the French window wide open too!

Being totally convinced that I had disturbed the would-be burglar each time, we took great care to put over the steel bar across the wooden shutters and then I made Helena check that I had done them properly. The caper went on for quite a long time and although there had been a number of burglaries in St Leonard's Road, we never lost a single thing. It was all very strange.

I decided to change my tactics. Instead of creeping downstairs in the dark, I would snap on the lights and, at the same time, charge down the stairs holding the French bayonet, which was normally kept under the bed. I would point it straight out in front of me, letting out one hell of a shout as I went, but the results were just the same.

Fluffy sometimes also knew that something was odd and would arch his back and look really frightened, even though he was not of a nervous disposition. Incidentally, Fluffy originally came to us one winter's night looking as if he was about to expire on the step of our back door. To say that he was painfully thin would be an understatement. We had never seen such a sorry creature in all our lives.

It was quite miraculous that he survived, but several saucers of milk laced with brandy obviously helped to stave off his imminent demise, but it took a very dedicated Helena to nurse him back to good health and live with us for the next 27 years. We believe that he was already 2 to 3 years of age when he crawled in to die.

Daphne was generally good, but as I said, at times we found her moody, although this was a very small irritation to cope with when compared to her hard work and natural liking for Rosalyn. I had been very busy at the factory and what with the numerous trips abroad, time seemed to go by in a flash; suddenly we all knew that Helena's baby was due very soon, particularly if her size was anything to go by! I have to admit that, frankly, I did not have the courage to own up as the person responsible for her enormous condition… It was something to behold! We only went out at night and then I would walk several steps ahead in order to ensure that no one was about.

Helena never complained. Not even once. Apart from the odd jimjams, she looked positively radiant and eternally happy. She really enjoyed having babies, so much that you felt that it really was the most natural thing in life and that it mattered the most. Often when I stopped to think about it, I could not believe that she could be so happy after leading such a busy and interesting life.

When we went for walks and if it was likely to rain, she would wear my Burberry; a double-breasted military style raincoat, but she had to wear it as a single-breasted coat! Bill Bishop was an absolute mini-gutter… he would not let Helena go to the annual dinner dance, as he considered her to be too large and might have put other people off coming! This really did upset her and rightly so; it took a very long time before she could forgive Bill for his heartless attitude. What a shame.

Late afternoon on 13th April 1954, we rushed off to Mowbray Nursing Home. I drove there like a demon, as I had been told that second babies were born within the hour when labour pains became regular and at short intervals. I was simply panicked and went through the Heavitree built-up area again like a scalded cat.

Unfortunately, Helena was suffering at the time from a rather severe attack of hay fever, which would often turn into a serious attack of asthma. I was very worried and tried to make the matron understand that she would need all the medicines that I had brought with me. But she just brushed me aside and ignored all my supplications.

I went straight home. I wasn't sure whether I could have stayed with Helena or not, but I didn't even give it a thought. The very idea of seeing a gruesome display before my very eyes filled me with horror. I just could not stomach such a spectacle! I believe that somebody said the other day that Him up there made a design fault and should have gone back to the drawing board… I think that metric measurements got mixed up with imperial!

At seven o'clock in the evening, I decided to telephone Mowbray and much to my surprise, the matron came to the phone and said, 'Ah, yes, everything is fine Mr López. You have a lovely baby girl... And there is another!' Well, I could have been knocked down with a feather... I was dumbfounded.

'Did you say twins?' I stuttered. I dropped the phone and presently I heard a voice saying, 'Are you there Mr López?'

'Matron, are you sure that you are talking to the right person?'

'Most certainly I am, you can come and see 'them' as soon as you like' she said.

★

Although Charlie was generally a very calm person, even he was somewhat taken aback by the news. But in no time at all, he was not only recovered but promptly poured out two large whiskeys for the both of us. I couldn't wait to get to the nursing home, but without even a bunch of flowers or a box of chocolates... Done it again! I just hope that I have made up for it since, but I suppose that it was such a mortal sin, that it would be asking too much to be forgiven. After all, how often does one have twins?

To be absolutely honest, I did think about some flowers when I saw other new daddies walking in with huge bunches of them, but alas, too late for me to turn back. Luckily, I am such a hopeless person that Helena has always loved me, in spite of my many shortcomings and frequent lapses of memory. It has got so bad now that she repeats everything as a matter of routine.

With eager anticipation, I rushed into the ward, nearly sending one of the nurses flying. And there she was, the cleverest and bravest of them all. In spite of the mammoth effort, Helena looked quietly serene. After having given her many hugs and kisses, I did have the presence of mind to thank her for all she had done. I saw that she gave a couple of glances around me, but I pretended not to have noticed her search for the expected flowers.

Although it was only an hour or so since I had learnt that we had become the proud parents of twin girls, I was already getting used to the idea. By the time I had got to Mowbray, I felt quite an old hand! I thought what fun it was going to be... Well, we had thoughts of having four children originally, but now that they were doubling up on us, I thought that we should practise a little restraint!

I could not contain my excitement any longer. 'Can I see them now?' I said. But Helena casually and very calmly said, 'One of the babies was very dark.' Oh my God, I thought to myself. Grandfather had been commander-in-chief in Cuba during the war with America. Had he been a bit too friendly with the natives? Perhaps fraternising with some of the local talent? Perhaps it was a throwback? Presently, our twin girls were wheeled in by two nurses. They were beautiful… The twins, of course! I didn't dare look, but after a searching glance from one to the other, I could see that they were both absolutely gorgeous, not a bit dark.

'Ah you see, Mr López, one had to have a blood transfusion,' said one of the nurses, pointing to the baby yet to be named Teresa. She apparently had been a breach baby. Now, this is very technical and I will explain it for the uninitiated. It means that baby Teresa came out feet first. I must say I was beginning to wonder why our doctor had not been able to spot that Helena was about to have twins. Well, it would seem that in those days, a doctor could not tell whether you were having twins if the two hearts were over each other. You could only hear one heartbeat. As x-rays were frowned on, you had no idea how many babies were on the way. To be somewhat technical again, you got identical twins when they shared the same placenta, hence our two cherubs being identical. Dr Harvey could be forgiven for the lack of warning.

I gathered that Helena's first reaction to being told that she had given birth to identical twins was not exactly one of joyous celebration, as she was not overpleased at having spent a great deal of money renovating Rosalyn's Silver Cross pram. But the chagrin at having wasted our meagre resources soon gave way to the pleasure of looking forward to the future.

The homecoming was thrilling as our lives were never to be the same again. Charlie was over the moon; an acting grandfather to so many children was more than he could have wished for. He said that he had got himself a family without having to go through all the paraphernalia of having to get married first!

Daphne couldn't keep her hands off the twins, each in a separate carrycot. But it was Rosalyn who was ecstatic at having two dolls to play with. She found it difficult to realise that they were real. There was nothing for it; we had to buy another pram. Silver Cross, without a doubt, but this time a double-sized one with the seats in line, fore and aft. This Silver Cross had even bigger wheels than our first one!

Rosalyn began to realise that she now had two sisters, not one as she had been told to expect. I suspected that she was also feeling a little jealous, as she was still only 18 months old and now not getting quite so much attention as before. But the twins were, after all, living dolls and a lot of fun. She would insist on sitting in the middle of the pram and to be dressed like the twins. If anyone said 'Oh, what lovely twins!', she would pipe up and say that she too was a twin! The twins lay at each end of the pram with the two hoods partly up and Rosalyn sat sideways on in the middle with her feet in the central well of the pram. They were all quite a picture in their most elegant big pram, which had these enormous wheels. The whole pram was in black with silver chrome fittings and silver lining along the sides.

Chapter Fourteen

Business expanded rapidly. We went into radio frequency welding of flexible plastics and as we did not have sufficient room for all the new machinery, we leased a factory in Marsh Barton Trading Estate, where we were to concentrate all plastics production. The weekend before we were about to move and install the plant, a large part of Marsh Barton became flooded.

Our new factory was now four feet under water and all the delicate electrical equipment was ruined as a result. We had to find temporary accommodation urgently. We finally settled for the Old Police Station in Waterbeer Street, next to the Mountain Café behind the old Woolworths. The smell from the café could only be described as dreadful, especially when they cooked cabbage.

Some of our machinery was placed directly over the café, where the floorboards were rotting, and consequently, the ceiling cracked under the strain and showers of dust fell into the cooking. Rats and mice were everywhere, but in spite of the unfavourable working conditions, we were able to get into production in a very short space of time.

At tea breaks we would all wander over to the old sprawling building which originally housed the main Old Police Station and also the headquarters for the fire brigade. Hence, we found many ancient uniforms; solid brass fireman's helmets as well as old police gear including handcuffs and truncheons, a load of antique firearms and firefighting equipment. The old prison cells we converted into temporary stores so that we could lock up. The building was in an awful state of repair from the effects of the Luftwaffe blitz on Exeter. You would require an old street map of Exeter to locate the site of some of these old buildings in the centre of Exeter, as the majority were pulled down to make room for the redevelopment.

★

Although Daphne proved to be a tower of strength when Helena and the twins came home, she became progressively more sullen and miserable as the novelty of having the twins wore off. It was no surprise to us therefore when, one day, we found a note from her to say that she had left for good and gone to live with some relatives in London. Daphne's sudden departure naturally caused a major problem for Helena, whose life was already frantic enough without Daphne leaving her in the lurch.

What with Charlie, me and the three girls, who were still only babies, plus a student from the International School, Helena was always kept busy, but also by that point, absolutely exhausted. Many a night I found her in tears lying on our bed, too tired to even undress.

Something had to be done. However, it was not easy to find suitable home help and for the next few months we had a string of girls living in, each one worse than the previous one; it was one disaster after another and then… a miracle! Mavis arrived and life took a turn for the better. Not only did she work like a thing possessed, but she was always bright as a button and went about doing her numerous jobs with a breezy manner, making light of everything. In fact, it was a pleasure to have her in the house and we all took to her and treated her like one of the family.

Mavis only had one little habit, which took a while to break. She came from what was then darkest Dunsford; a no-man's-land part of Devon, just south of Exeter. Her father was a farm labourer and their tithe cottage had no mains services or drainage, which went a long way to explain why she always left water in the basin after she had washed… kept for the next person to use. She had been taught to conserve the precious liquid.

One evening whilst working on at the factory, I had a desperate phone call from Helena.

'Come at once. I want help urgently,' she said and put the telephone down without giving me an explanation. Naturally I dropped everything and went home post haste with my colleague, Hopely, who ran the plastics section, as things sounded pretty dire. When we arrived, we found Mavis in a state of shock and Helena sitting at the bottom of the stairs in a flood of tears.

The three girls slept in one great room, as Rosalyn wanted to be with the twins for company. When we opened the bedroom door, we found Rosalyn tied to the fireplace and each of the twins tied to the sides of their cots. It was obviously the only thing that Helena and Mavis could think of at a moment of sheer crisis... The screaming was enough to wake the dead.

Looking around I could see that the children's frieze, which we had only just put round the walls, had been torn to shreds, excrement was smeared on each of them, on the walls, on the paint work and on their cots. One glass pane was missing from the big sash window and there was glass strewn across the floor. It was an unbelievable sight. We were speechless... We stood there with our mouths wide open unable to comprehend it. At last the yelling stopped. The girls had frightened themselves into a silence as they too stared at the four grown-ups looking as if they had been turned to stone.

All three had been promised a good hiding when I came home. At last I spoke.

'Come on now. I'm sure that you didn't mean to do it. Go and give mummy a big hug and kiss and say sorry.' I wasn't sure what Helena's reaction would be, but I need not have worried as she hugged all three together. We untied their walking reins and the three trooped naked into the bathroom and got into the bath together. We all set to, and each was given a thorough scrub, new pyjamas and a hot drink, and whilst all three sat on the table by the bathroom window, we cleared up the bedroom as best we could. Finally, when they were all tucked up, looking like little angels, each began to speak at once.

'It was her fault!' they all shouted. But we were pretty sure that Rosalyn had been the ringleader. She got Teresa to hang onto one of the long window curtains as she pulled her back and then let her go against one of the small windowpanes. It was fortunate that she did not go right through the window, probably because Teresa had not taken off her nappies, otherwise it would have been a real disaster. The window was directly over the basement backdoor... All of a 25-foot drop.

After such a traumatic episode, all other family incidents paled into insignificance. All I can remember is a period of domestic bliss... Perhaps as I was not quite so immersed in the household everyday events, as my business was expanding rapidly and I found myself having to work on at the office more and more, late evenings and even most Saturday mornings. On top of all of that, the business required frequent trips abroad.

The net result of my enforced detachment from the family meant that Helena was very much on her own to bring up the girls, with the devoted help of Mavis. It had been impossible to consider a holiday. In fact, Rosalyn was three and a half years of age and the twins two years old, before we managed our first holiday. We rented a lovely little chalet at Pentewan, near St Austell, where the sea was a beautiful aquamarine shade, due to the china clay workings nearby.

I cannot remember the reasons why, but Mavis did not come with us to Cornwall. She stayed behind and looked after Charlie, for which I discovered later, he gave her a beautiful brooch. Mind you, Mavis could never have got into the car... We had taken everything bar the kitchen sink, but it was a close thing. The car was jammed to the gunnels and the girls had to sit on a plank of wood, which I had squeezed between the sides of the car and placed on top of the leather armrest on either side of the rear seat. That enabled the three to see out, thus reducing the level of internal strife.

The journey proved quite eventful and our worst fears were to be realised when we got to the sharp bend and steep hill at Callington... Our poor old Daimler could not manage it. Helena and the three girls had to get out in order to lighten the load. We had to stop once more to administer a severe warning, but in reality, I had to stop and adjust some of the luggage and bundles on the roof and at the rear, where we had tied the three potties and God knows what else!

No sooner had I finished securing everything, when all three wanted to have a wee-wee. I was certainly not going to unfasten three potties. Oh yes! They would never wait for each other to use the same potty. So for their first time in their lives, to my knowledge, they all took down their knickers and enjoyed crouching by the roadside in a row, waving to passing cars... This initiation was to prove helpful some years later in Spain.

When we eventually got to Pentewan, we found that the chalets were very modern and easy to run. Although built quite close to each other, they were placed in such a way that you could sit out on your patio without being overlooked. We were lucky to have friendly people on either side of us. One lot came from Canvey Island and the other from some part of Wales. Both had children, so our lot soon became immersed with them, leaving Helena and me to spend more time together... Something of a novelty.

As the chalets were virtually built on the beach, we did not have far to go for our bathing. The girls were only allowed to go when they had supervision. It was a lovely holiday. However, at the start of the third week, we were disappointed to find that Charlie had booked himself into a B&B in the village, only a short walk away. I was anxious to give Helena a break, but in the end it turned out all right, and he too enjoyed the change.

The weather for the whole of the holiday was absolutely wonderful, and hence we spent most of the time on the beach, but we did find time to explore the area. We loved Carlyon Bay with the glorious show of hydrangeas, particularly close to Carlyon Bay Hotel. We also liked Charlestown and Mevagissey. In fact, we made up our minds to make that part of Cornwall a priority for the next visit.

Much as we loved our very first holiday, the effort of having to prepare for it again was more than we could face, so when we found out that there was a cottage to let, during one of our regular trips to Abbotsham in North Devon, we jumped at it and leased it for a minimum of five years, with the option to renew for just a few shillings per week. Abbotsham is near Bideford, a small port which I knew well having spent my early years in England nearby, between Northam and Westward Ho! The cottage could not have been more primitive. It was gorgeously simple and we had great fun furnishing it. Certainly, it was a sharp contrast to our elegant Georgian house in St Leonard's. The children enjoyed living rough for a change. I believe Helena partly liked it for that reason too. We could all let our hair down without having to worry about it, or keep it neat and tidy.

Cosy Cottage in Pump Lane had one bedroom, but it was large enough to take two double bunk beds. The landing at the top of the stairs was so wide that we were able to fit in a double bed. It was a very convenient arrangement as it allowed the children to go to bed first in the only bedroom and we could follow to our very comfortable double bed when we liked, without disturbing the girls. The toilet was at the end of the garden. Fortunately there was a good path to it, but a torch at night was absolutely essential. Suffice to say it had an excellent flush, which was a positive plus!

The cottage was in the main street and the last building on our side; then you had about a three-quarter mile walk to the pebble beach, which was generally deserted, even at the height of summer. In winter, the gales were quite horrendous. Driving rain made it necessary to wear sou'westers most of the time, but we found it very exhilarating when we returned to a warm cosy log and coal fire in our front room at the cottage. The ferocity of the winds explains why the few rather mangy, stunted trees all lean away from the shoreline.

The cottage was suitable all year round and so we made friends with several village folk, including the village blacksmith and his wife. They lived on the other side of the road. As we had no bathroom in the cottage, we arranged to have a bath in their house for a shilling at a time. It became a pretty common sight to see us traipsing across the road in the evenings in our dressing gowns. In the summer it wasn't a problem, but in the winter it was very different. You nipped smartly across with a howling gale blowing up your whatsits!

The village blacksmith had endured poor health for years and was not expected to survive each winter, according to general opinion, and this had been so from time immemorial. Surprisingly, however, he was always there in the spring and everybody was jolly pleased and congratulated him on it. It was therefore somewhat surprising that he continued to shoe horses and mend farm equipment. We were most concerned, as he looked as if he would not make it to the next morning!

We were woken up very early one Saturday – it was the morning for our regular bath. A lot of coming and goings were occurring across the road at the blacksmith's house. We recognised the doctor's car parked outside and naturally assumed that the expected worst had happened at last.

You can imagine our astonishment when we learnt that it was the wife, a tall robust woman who had died in the early hours of the morning, from a heart attack. The whole village was stunned as she had been looked upon as a tower of strength. Well, it had changed things for us immediately. No bath that morning or any other morning.

Bathing in the sea was now the only alternative open to us. The sea was fine in the summer months but a different story in the winter. We tried it just once, when there was frost in our back garden. Never again. So we made do with a sponge down in the kitchen. No so much fun, and consequently, washing became less frequent.

We struck up friendships with the landlord and his wife at the New Inn in Abbotsham, where we had our own room... Well, we treated it as ours! It was where we spent most winter evenings. What else was there to do? I became quite addicted to Ushers' bitter, but as we did so much walking I didn't get a beer gut, and it certainly kept the cold out. Our room at the pub was really the off-licence, so we did get the odd person coming in for bottled bitter and stout. There were no cans in those days!

We became good friends with Consi and Martha and often arranged to meet in the pub together with their children. Martha was a nurse and consequently kept rather odd hours. Fortunately, their daughter liked looking after our girls. She was a very pretty girl and her young brother was a happy little soul who went about humming a TV sales ditty... Pomp, pomp, pomp... pomp... Esso Blue! It was a paraffin advert.

We had many happy hours at the cottage, at the New Inn, walking in the village and the long walk down to the beach which, as I said before, was nearly always deserted. It was best when the tide went out as we usually went prawning for our supper, in and out of the rock pools. One pool in particular was very deep and full of all sorts of sea creatures. One day we caught two large lobsters in different pools. One was very easy to catch as it was stranded and didn't have enough room to make a fight of it, but the other was a totally different kettle of fish, so to speak! Helena had spotted a real whopper, the biggest I have ever seen alive, in a crevasse at the side of a very deep pool. Every time we tried to get him with my gaff, he would craftily slide further back into the crevasse.

We had brought a picnic lunch onto the pebble ridge and Helena found a sardine sandwich that had been left over, which, much to my amazement, she offered to our rather shy lobster! I thought what a daft thing to do, but no! The temptation was more than any self-respecting lobster could resist and out he came, bold as brass and in a trice, I had him dangling from my gaff! We weighed him and found that he was just over two and a quarter pounds. The next job was not so good however... we had to boil him alive. I can't remember from whom we borrowed the huge pot, but I clearly remember putting him into the boiling water. It was not the nicest thing that Helena and I have ever had to do... but it did taste wonderful.

★

Time went by very quickly at No. 1 St Leonard's Place; it was such a happy place, and our ghost continued to lodge with us. He would do just the odd haunt or two in order to keep his hand in, sort of paying his way, you might say. He was friendly at all times and we never had cause to take any drastic action. In fact, it was a pleasure to have him in the place. Well, he never let on as to his sex, but we were pretty sure that it was a he.

From the day that we were married, we kept in close contact with St Leonard's Church and the very snooty vicar that married us and also christened our three girls, Reverend Royds. A teetotal man with very strong convictions; we did not realise how strict he was, but we were to find out one afternoon when he came to tea. We were sitting by the fire in the beautiful sitting room, when the door opened and our three girls, looking angelic, came in with Mavis. They had just had a bath and were ready for bed, but because the vicar had held up me and Helena over tea, and was in fact about to have another cup... Rosalyn, our eldest, who always acted as spokesman, said in a very injured voice 'We haven't had our whisky, Mummy, and we are all ready for bed!'

Mr Royds jumped up and left immediately; so quickly in fact that we were left unable to speak. We would have liked to explain that we always put a tablespoon of whisky in their hot milk at night because of their troublesome teeth, plus it also helped them to go off to sleep. We never saw our vicar again, but it was no great loss as he was overbearingly pompous!

It was soon after the episode with the vicar that Rosalyn gave us the fright of our lives. We could not find her after lunch one afternoon. We searched everywhere, but the St Leonard's house was very large, including a huge basement with a network of rooms where, in earlier days, the servants would have lived. For this reason it took us quite a time to discover that she was in fact not in the house.

We decided that she had gone out onto the main road. We were able to see up St Leonard's a long way, so we thought she must have gone down to the very busy Topsham Road. Helena by now was frantic; we all were. To our momentary relief and then utter horror, we saw Rosalyn halfway across Topsham Road, with cars having to stop suddenly from either direction, leaving a path for her to complete her crossing to the opposite pavement.

Once home again, 150 yards or so, the debriefing onslaught started, but not before several wallops had been administered to her bottom out of sheer relief at getting her back home safely. It appeared that there had been a bit of a tiff between Rosalyn and the twins, so Rosalyn decided that three was a crowd and that the only answer was for her to go to sea… She had packed her teddy and her constant companion, Brownie, who was her very worn rag doll, plus some sweets and her Noddy book, entitled *Noddy Goes to Sea*, which I had been reading to them in bed each morning.

★

Life from then on continued without any more major upsets. It was by and large pretty uneventful, just a typical, very ordinary family, bringing up three little girls during their early formative years. Each day would seem much like another; lots of picnics and visits to the seaside, Exmouth and Dawlish Warren being the most popular and, of course, very near to us.

Charlie retired from the electricity board soon after his sixty-fifth birthday and was able to concentrate on the building of his model railway. I believe that it was the largest '00' scale in the country. It occupied the two largest basement rooms in the house. They were interconnected by a long tunnel through the very thick dividing wall. I did not have a lot of spare time, but I did manage to make most of the station buildings, much of the scenery along the twin tracks, the embankments and the countryside in between. However, I was carefully supervised by Charlie, for he was meticulous about detail and the correctness of scale.

Charlie invented the two rail electric system for model railways initially for '00' gauge, but later was introduced to other sizes by Bassett Lowkes of Northampton, to whom he sold his patent for a nominal figure. His system avoided having to have the third rail to carry the current. Why he sold the patent so cheaply I have no idea, because Bassett Lowkes went on to make a fortune afterwards. The principle was very simple: the wheels of all the rolling stock were insulated from their axles and the current was picked up by brass spring contacts on the rim of two of the engine wheels. Both rails were live and there was no need for the third rail.

Charlie became quite famous amongst the railway modelmaking fraternity as well as the legion of railway enthusiasts. He wrote articles regularly in the *Model Railway News* under the pseudonym 'GER'. He completed the major part of his 52-lever interlocking signal frame whilst in Spain, but it was not until the early 1950s that he managed to install it into the composite system. It was quite uncanny to see up to eight trains running simultaneously, and if the operator failed to clear the line, the two trains would stop if they had been on a collision course.

We had many visitors over the years as Charlie's frequent articles in the *Model Railway News* often contained photographs of various sections of his own comprehensive and intricate track layout in the house at St Leonard's. I remember one such photograph in particular. It was of the signal gantry at the approach to the mainline station, which was a replica at that time of the real thing at Clapham Junction. The semaphore signalling arms had individual miniature electric magnets under each signal post, but were well camouflaged in between the girders making up the gantry bridge. The whole layout was built on permanent trestles with a skirting board on either side, so that all the electrics were hidden from view. I believe that it was the best set-up in the whole country, which is why so many came to see it!

Chapter Fifteen

We were quite glad to have lived so close to the Royal Devon and Exeter Hospital as we had to make quite a number of unscheduled visits. You would think that little girls would never get hurt. It was Teresa that seemed to get the worst of it; once she fell back on her bottom and caught her foot underneath herself in such a way that the prong of the buckle on her leather sandal stuck deeply into her bottom. On another occasion, by accident, Rosie stuck a pencil into the side of Teresa's left eye. It looked awful, but within a few days all was well – not even a mark!

Before we had central heating installed, Helena used to light a paraffin stove in the bedroom just before the three girls had their afternoon siesta. She would leave the stove on a low flame if it was very cold out. On one occasion, for some reason or other, she decided to check the girls a little earlier than usual... Horror of horrors, the room was full of thick black smoke. The Aladdin stove had become sooted up and as the girls were fast asleep, they would have died from carbon monoxide fumes poisoning. It was a miraculous escape.

One visit to the hospital was for all three at the same time. Their navels had protruded from birth and the surgeon, Mr Denddie Moore, decided that it would be better to have all three in for surgery together so that they would keep each other company and be less frightened. This was fine for them, but their poor mother and I nearly went frantic. They had never been out of Helena's sight before and now all three were having operations at the same time! Mr Moore, who had children of his own and whom we knew well, realised how worried Helena would be so when he finished operating on the last of the girls, he rang Helena personally in order to reassure her.

We always had two birthday parties each year; 13th April and 13th October, so our three girls celebrated a birthday and a half a birthday during each year. As I was working, I usually came home as the tea finished but on one occasion Mr Moore's youngest daughter, Joanna, was most unhappy and after much attention, she informed my wife that it was not a proper tea as there weren't any ham sandwiches!

We had another panicked event when the central heating was being installed. The engineers had turned off the gas in order to connect the huge Pillinger boiler, so Helena decided to cook brunch; eggs, bacon, sausages, baked beans, tomatoes and fried bread on the paraffin stove in the middle of the kitchen floor. Most of the floorboards were up in order to lay the pipes. It was unfortunate that Helena had chosen to put her stove down on some floor boards that had not been nailed down... Suddenly one end came up and all the boiling fat spilled over Helena's right arm. Brunch ended up in the bowels of the house. In a flash, we were at the RD&E again and it was serious. Helena's arm had developed an enormous skin blister, from her wrist to her elbow and it was full of fluid. It was quite frightening, but there was not a squeak from her; she was really quite stoical.

★

All three of the girls had sandpits in front of each basement window at the front of the house, under the veranda. It served as a house for each one and they spent many hours playing with the sand or making believe that they were running a home like their mother, with the added advantage that they could see people going by the front gate. Although I had made an attractive and practical playroom, complete with a blackboard built into one of the walls and easy-access cupboards for their enormous quantity of toys (most from Charlie), we found that the three girls spent much of their time in the basement, for the railway track really did prove to be such an attraction. When we finally finished installing the huge Pillinger oil-fired boiler in one of the basement rooms near the railway track, we decided that it would be more sensible for the three girls to have most of their toys in the boiler room too, as it had built-in shelves and a window overlooking the front lawn.

A young Nigerian and his wife came to live nearby. He was a

tutor at Exeter University and they had a lovely little baby girl called Kesia. She was jet black and our girls became fascinated by the colour of her skin. In those days there were no black people in Exeter, and certainly not in St Leonard's, so our girls were determined to get to know the mother so that they could handle the baby. The mother and father were charming people and were only too pleased to let our girls play with Kesia. Many times when the girls thought no one was looking, they would try to rub off the black colour from the baby's arms and the back of her hands.

In no time at all, the three large dolls had to be discarded and we had to buy three new black dolls... All called Kesia, of course! Each was dressed differently so there were no problems in knowing whose was whose.

Auntie Harvey also lived close by and she became very fond of the children. She had been a schoolmistress and knew how to interest them. At that time, Rosalyn had developed an enormous wart on her hand – it was quite nasty and it worried Rosalyn no end. One day, Auntie Harvey decided to do something about it. We do not know what she did or what she said, but the wart disappeared and Rosalyn was sworn to secrecy. To this day, we still do not know, as Rosalyn has kept to her side of the bargain... It is still a secret!

St Leonard's was close to the centre of Exeter, so it was quite easy to walk in to work or to the village where there were many good shops and an excellent pub at the end of the road. When I had a Masonic function in the city, I would walk there and back so that I could enjoy having drinks with my friends, especially at the Exeter and County Club in Southernhay West. I remember one night after a Lopes Lodge meeting and dinner at the Royal Clarence Hotel, we went down to the club as usual to play snooker at a bottle of scotch per corner, which we proceeded to drink up between each game! I decided to stay at the club that night as walking home would have been somewhat difficult!

I will avoid mentioning names, but one of the party decided to drive his car home. His car, however, was parked right in the middle of the car park in front of the club. No matter how much we tried to dissuade him, he was determined to get his car out, and he did! But not before several cars had suffered some nasty bumps. Having got out onto the main road, we waved him home with some apprehension and were about to rush into the club as it was tipping it down with rain, when we heard a loud noise. We rushed out and found our friend having an argument with a No Parking sign outside the RD&E hospital. Within seconds, there was a copper leaning down and talking to our friend through his opened passenger window and we heard him say, 'Evening sir... What a damn silly place to put a signpost... Trust you are all right, sir... Goodnight sir!' The copper obviously knew him.

The business continued to expand at a pace and we became major suppliers to Marks & Spencer, Austin Reed and Debenhams, to mention but a few. Later we designed and manufactured for Burberrys, Aquascutum and other well-known companies. I became pretty well known in the luggage and plastics field due to my numerous successful designs and patents. I remember one patent in particular, as it made quite a lot of money!

Marks & Spencer were having a great run with one of their baby umbrellas... The problem was that they could not get enough crook handles from Italy. Apparently, the little crooked steel handles had to be covered with an embossed PVC sleeve, which was a complicated and difficult job to complete that the Italians had had done by outworkers, who could make one every five minutes. The senior product design manager at Marks was asked by his immediate director to source someone in the UK to augment the Italian supply of handles as they were losing sales and we were luckily having one of our typical wet summers! Luckily for me he asked if I could help, and by a stroke of good fortune I came up with a production method which was capable of producing five every minute... Twenty-five times faster than the Italians. I got orders for one and a half million to start with. As they were delivered in good time, more orders followed.

I charged the same price as the Italians; I was very happy and so were Marks... To start with. But when they realised that I was making a bit of a killing, they tried to get me to bring the price down. At first I was inclined to agree, as it would mean that Marks could in turn reduce their retail price, but I discovered that they had no such intention... They considered that the product was retailed at the appropriate price so I refused. They stopped sending me orders for a whole month, thinking that I would cave in, but I held out and hoped that the wet summer would continue, which it did! New orders followed, up to three million in all and all at my original price!

Very soon after my little tussle with M&S, I made a soft folding wardrobe case in which one could hang a suit on a special locking coat hanger, under the name of a Pac a Nite. I made the size so that it could be carried on an aircraft as hand luggage. It was an instant success. Shell introduced it to their executives as it saved time by avoiding them having to pick up their cases from the carousel luggage reclaim area. I had numerous write-ups in the *Financial Times* and the *Daily Express*. Selfridges ran it as a mail order item in the *Sunday Express* for a couple of years. As I had taken out a patent, I had it to myself for many years. Obviously, I manufactured many models on the same theme. However, the original carry-on model was designed to fit behind one's legs when in a sitting position, as the lockers in the earlier jets were very small. The company, which I was to buy at a later date, was called Gilchrist and Fisher Ltd, but we decided that as we owned the Kingfisher trademark for luggage and ancillary products, that we should form a subsidiary company to include the trademark; Kingfisher Luggage Ltd. This proved a great success and we became a well-known and respected company in travel goods.

As things were going pretty well in my work, we continued to improve our home and that meant that some of our earlier furniture had to go. The children too had accumulated an enormous collection of toys, a portion of which never saw the light of day. We decided to have a sale at the house and as Helena had been the instigator of the project, she was put in charge. Well! She was a natural and organised everything to a T. The most tempting advert appeared in the afternoon edition of the *Express & Echo*, which was basically then an Exeter evening paper. I decided to go home a bit earlier that afternoon in order to give Helena and Mavis a hand.

I found that I couldn't get the car into the drive as it was completely jammed, so I parked in the main road and walked up the drive towards the front door. There was a queue from the hall to the garden gate. I attempted to pass by but I was ordered to join the back of the queue, in no uncertain terms! When I explained that I was the owner, they let me through and said, 'Sorry, governor' as I passed. As I got to the front door, a chap came out holding up a small oak dressing table, which I recognised as having come from Charlie's bedroom.

'Hey!' I said. 'That's not for sale.'

'Oh yes it is. Your missus said I could have it,' replied the man and disappeared in haste. What Charlie would say when he got home, I shuddered to think.

The sale had been planned just before Christmas as Helena felt that the toys would be a big attraction. How right she was. They were snapped up – even fought over, and there were quite a few ugly scenes. Everything went and more besides. Fortunately Charlie was very good about his old dressing table and said that he never really liked it anyway. We all sat down for a well-earned drink, although Helena had done the lion's share of the job. We drunk a toast to her… This was a side to Helena that had been previously unknown to me. We all looked up when we suddenly heard a terrible noise. Someone was banging on a door… It was coming from the basement. Mavis had locked it after the sale. On opening the door we found one of the customers, clasping a doll's pram and waving some pound notes in her fist!

Despite the sale, the girls still had a heap of toys; they now also had doll's prams designed as the large Silver Cross was. The prams were so large that they could take a real baby in them. The three girls made quite an impression wherever they went, as they were outstandingly pretty and, as they were also so close to each other in age, they were often taken to be triplets. They were in fact born exactly 18 months apart.

Because life was becoming a little easier, not only physically but also financially, with the foreign students being a big contributor, we found that we could afford to go on holiday more often, Burgh Island being one of our favourite destinations. When we went for the first time, the whole island and a large part of the mainland opposite belonged to the engineering company Guest, Keen & Nettleford, who built it for their directors. It was a lovely place and so free and easy.

We always had a suite on the first floor which incorporated a large circular terrace facing the mainland. The suite consisted of three bedrooms, a sitting room, a kitchenette and a very luxurious bathroom. The hotel expected you to prepare or go out for all meals except dinner, which was always a grand affair. We all put on our best bib and tucker. The food was excellent and after dinner we would dance in the ballroom or on some evenings, we had horse racing and gambled on each race. The horses were plywood cut-outs which travelled on strings across the whole length of the ballroom... Someone had to wind a handle at the end of each lane to make the horse move forward and we all took turns!

You could walk around the whole of the island, which consisted of about 26 acres. It had a natural pool made by the formation of the rocks; the sea could only come in over a man-made breakwater. As the tide receded, it left enough water trapped to make a large deep pool suitable for swimming and a sandy shingle beach at one end. Beyond, there was a concrete jetty on which there were tables and chairs and behind that, a few steps up, there were changing rooms with a viewing balcony above.

As you arrived at the island when the tide was in, having just got off the Sea Horse, also known as the Sea Tractor, you climbed up a steep path wide enough to take a Land Rover to the hotel entrance. But before you arrived at the hotel itself, you passed the fourteenth-century pub called the Pilchard Inn, which looked exactly as you would imagine a smugglers' den to look like. From all the accounts that we heard over the years, it certainly had been a den of iniquity!

The hotel was very airy and spacious and built by Archibold Nettleford of GKN Industries, in about 1929, without regard to cost. Many celebrities stayed there in the 1930s, such as the Prince of Wales and Mrs Simpson, Noel Coward and, of course, Agatha Christie, who wrote two of her best-known mysteries on Burgh Island; the novels *And Then There Were None* and *Evil Under the Sun* were both inspired during the many visits to the island by Christie, who considered it to be the Ritz of the West country.

The whole of the hotel is in the 1930s style. Opposite the entrance, there was the Palm Court, with its magnificent peacock dome where Harry Roy and Geraldo entertained. You entered this enormous circular glass-domed room by marble steps from the entrance hall down to the main floor, which had a fountain to one side and more marble steps either side to a higher level. This level had globe lights on slim posts dotted about, as well as palm trees. On the lower level, there was a cocktail bar where you could most definitely not get a pint of beer... Far too common!

The dining room had windows facing the shore and at one end, there was a further bar and coffee area looking out across the island through windows either side of a most beautiful figurehead of a woman leading the prow of the ship. We had every amenity that we desired. Even a *camera obscura* was located on the highest point at the centre of the island. By the slipway, close to the entrance of the island, there was even a pirate ship, The Jolly Roger, complete with the black flag displaying the skull and crossbones, fluttering at the mast. The ship was lying on shallow rocks and it accommodated a family on holiday. But as time went on, we noticed that it was getting rather battered by the winter storms and it ultimately broke up.

All holidaymakers who came by car reported to the huge undercover garage on the mainland, directly opposite the island. You unloaded the luggage onto a Land Rover and your car was parked, protected from the elements, for the rest of your holiday. If the tide was in, the Land Rover drove the holidaymakers down to the Sea Horse, which we had heard about and often seen pictures of in holiday brochures. It was really exciting... You had to climb up some steep iron ladders onto a platform built over a gigantic tractor with 8' wheels at the rear. The driver sat in front of the passengers and drove the Sea Horse by a long steering column, which went to the front wheels that were always under the sea. Our vehicle had a canopy like a surrey. At high tide the sea came up to the top of the wheels, just a foot or so from our feet! If there was a spring tide, there were no trips to the mainland and no trippers crossing over to the island for the Pilchard Inn!

It was on such a crossing that we were amazed to find Helena's ex-fiancé frantically waving to us as we were scampering down the gangway of the Sea Horse. He was with his wife and children and they were going to stay in the hotel for the same period as us. At first we thought that it might be a bit embarrassing as he was such an extrovert, but not a bit of it! He was to prove good fun and the life and soul of the party... Of which there were many! Being an auctioneer, he took over the horse racing and ran a book. Just to add to the jollifications, he would blow a hunting horn at the start and finish of each race.

When the tide was in and the Sea Horse was parked up the slope on the island, we were happily marooned, no more so than at night. The licensing laws of the mainland were temporarily dispensed with and the wooden shutters of the Pilchard Inn were closely sealed so that no lights could be visible from the mainland.

There were occasions when the landlord of the Pilchard Inn wanted to have an early night but he found it difficult to close, as the hotel residents made his life difficult. And of course, the money was good. One night, he and his wife had had enough and shouted last orders at about midnight. I offered to take over the bar with the help of a couple of the other guests. Much to our surprise, he accepted, emptied the till and went!

It turned out to be a super evening, for when the other guests in the hotel heard what was happening, they too came down to the Pilchard Inn *en masse*. We couldn't work out the different prices so everything was charged at half a crown, no matter whether it was a pint, or a half a pint, or a whisky or whatever. We filled the till with money and at the end we had to use a cardboard box!

The landlord came in the next morning looking rather worried and had not slept a wink, as he had been worrying about the losses that we had incurred for him. When he saw the money overflowing from the till and the cardboard box of silver as well, he asked us if there was any chance of doing him a favour again!

One year we stayed in a chalet on the island. There were several dotted about on the edge of the cliff, usually up a small creek and built on stilts with ladders going down to the sandy beach. These chalets were great fun. When the tide was in, fishing from your own veranda was quite an idyllic pastime. The sound of the waves from under your bed at night helped to lull one off to sleep.

Next to us at our last chalet lived a very handsome retired merchant seaman by the name of Gerald. He must have retired in his thirties! He was most popular with all the females, no matter what age they were. I have a feeling that he lived in his semi-cave dwelling rent-free, perhaps for certain favours! I noticed that some nights he had the odd visitor; I never saw who it was, but the voice was not always the same.

On some evenings we had a barbecue by the pool and the young student girls wore bathing towels as sarongs or saris when they dished out the food. With lots of wine flowing, things got out of hand, particularly on moonlight nights – towels just would not stay on! Our girls, of course, had no idea what went on at night as they were safely tucked up in bed, and an alarm system ensured that we did not have to worry about them.

We still had our cottage at Abbotsham, so we stayed there many weekends and some of the children's school holidays in order to enjoy the primitive life. Part of the pleasure was the journey up to North Devon; Helena always made super streaky crispy bacon sandwiches with Branston Pickle... They were something else. We always stopped at the same place, on the middle of a bridge over a stream which had been by-passed by a new road improvement.

Chapter Sixteen

Over the years, we went through a series of puppies. First there was Whisky, a sheep dog. He had to go, as he was far too excitable. Then there was the jet-black poodle; he didn't last long either as he bit one of the girls. After our experiences, we decided to go without a dog for a while.

Helena continued to have students stay at the house, mostly foreign. We had one that we liked a lot. She was a schoolteacher and visited us at the cottage. Another was Mercedes, a Basque girl, very nice, and she too kept in touch with us for a while. After Mercedes, we had a very pretty blonde Belgian girl, but she didn't like having a bath and you could always tell when she passed by! Then there was a Swiss girl, who did bathe, but she wouldn't let Helena wash her clothes. When she had worn her knickers for a while, she cut the centres out and threw them in the bin!

Mavis by now had become part of the family, as well as a good friend to us all. The inevitable happened though, of course, as she was a lovely-looking girl. She got herself a boyfriend, but they were sensible and waited to get married until they could afford it. The wedding took place in Dunsford. Mavis's sister and our three girls were bridesmaids. It was a very pretty wedding and the marriage proved a great success. Mavis and Jeffrey raised a fine family and still keep in touch with us now. I am sure that Mavis had been very happy living with us, even if at times it had been a bit exhausting.

After Mavis left, we had a string of girls... We just could not find anyone to live up to Mavis's qualities. The first girl we tried bit her fingernails until they bled. Actually, she came to work for me in the factory some years later. We then had Christine, who worked well enough, but we never got to know her very well. The last girl to work for us was Brenda... She was a nice-looking girl, did not exactly set the Thames alight, but she was pleasant and was in the Salvation Army. I remember that she did not like going to bed without me checking under her bed with a torch, in case anyone was hiding underneath. I never made up my mind whether this was wishful thinking or not!

Although we enjoyed our fabulous holidays at Burgh Island and our frequent stays at the cottage, we were always happy to come back to St Leonard's, as we loved the house and felt very privileged at being able to live there. The garden was by no means large, but as I was so busy at the office, we employed Mr Bett, a lovely man, full of goodness and who simply loved the children, to look after the gardening for us.

One day he made quite a discovery. He decided to investigate an area of the garden which always appeared damp, so he dug the ground much deeper than usual, only to find an old cast-iron bath that had been buried. I gave him a hand to break it up with a sledge hammer. When we cleared all the broken pieces, he found a pewter plate and a scoop or measure, both inscribed with Roman letters and numerals. Exeter Museum was not exactly overjoyed at the find, but they said that the two items had belonged to a Roman soldier. St Leonard's area had been the site of a Roman camp. Soon after that, Mr Bett also found some solid silver sugar tongs nearby. We still have them.

Much as we loved our splendid house, there were definite stirrings for a move to the country. The girls hankered after ponies and fancied a rural life for a change. In a very short space of time, we found exactly what we were looking for; perhaps more correctly, what I thought everyone was looking for! I took the plunge and we bought a ranch-style house at Silverton, La Rosa, seven miles from Exeter. It was winter, but we sold St Leonard's within a few days of putting it on the market. In fact, we sold a huge proportion of the furniture, carpets and curtains together with Brenda. The new buyers had to leave Rhodesia in a hell of a hurry and left all their belongings behind. They asked Brenda if she would stay and she readily agreed so long as she could have her room exactly as it was. They agreed to pay for everything, including the ornaments and even the bed made up as it stood.

We arranged to move the day after Boxing Day 1962. Well, as it turned out, it was to be one of the coldest Christmases I could remember. On Boxing Day night, we had the biggest fall of snow for years! The removers arrived early in the morning, two great pantechnicons and four men. It was snowing in Exeter, but it had only started at about six in the morning and wasn't too bad at that point. Little did we know that it had been snowing all night in Silverton. We all got there, more by luck than judgement, but could not get up the hill as it was covered by a heap of snow – at least four to six inches in places. We had to start clearing enough of it to make a track for the furniture van and then it started to snow again. The removal men had had enough of it and decided to return to Exeter as the road might become impassable later. We were in a desperate situation.

The central heating oil tank was empty – what a mean thing to do! Charlie pleaded with the men to stay and unload and promised that if they finished, each one would get a bottle of whisky. That did it! I have never seen men move so fast in all my life!

The three girls were excited and just rushed about in the snow, although it was quite dangerous for you could not see any recesses or beds, as they were covered over. Then they came across the swimming pool. It was frozen over with a layer of fine, dry snow. Well, we could not stop them from putting on their bathing costumes... Goodness knows how they were able to find them! I could not believe my eyes. All three jumped onto the frozen pool and, as we expected, the ice gave way!

The screams I am sure could have been heard by the whole village, particularly as there were very few vehicles moving about. The girls stayed in the icy water until Helena insisted that they came out, as there was no hot water, to have a hearty rub-down! We found the switch for the immersion heater, but it took a long time to warm the water as the system had been switched off for ages.

It took all day to unload and place the furniture in position, and many boxes were left in the garage as we had no idea where some things were going; added to which, Helena was not exactly brimming over with enthusiasm, as it became obvious that she resented being yanked out of her beautiful home in St Leonard's. I was certainly not her favourite husband just at that precise moment!

The men proved absolutely marvellous and well deserved their bottle of whisky each. It was quite a problem to turn their big pantechnicon in our turning circle, as they kept sliding sideways on the hard, icy tarmac. At last they were away and we sat down and had a well-earned strong drink; the children too had their milk laced with a larger tot of whisky than usual. Everyone was exhausted and we all slept well in our hastily made beds.

Next morning, we found even more snow. It must have been at it all night. I decided in the late afternoon to clear sufficient snow from the drive so that I could go to work the following day. We all worked like Trojans and by dusk we had cleared enough to make it possible to drive down to the main road.

I got up early next morning, had a quick coffee and tried to get the car out of the garage. After much sliding about, I managed to get it facing down the lane. I could not believe my eyes – the end of the drive was completely sealed off by a huge mountain of snow! The snow plough had been down the High Street and had thrown the snow across the entrance to our drive. Well, that was that! It took nearly all day to clear it again, but at last I was away to work. I felt like a bit of a heel... I went and left them without heating and the whole house in one hell of a mess – not very kind!

Somehow I managed to get to the office in my Ford Executive, which had rear-wheel drive that wasn't ideal for snow or icy conditions. That day we had another heavy fall of snow and I found it impossible to drive the car, so I had no option but to stay with Dickie and Pops at Salmonpool Lane. I rang Helena in the evening to explain my predicament... Her comments were to the effect that it was all right for some!

I could not get heating oil delivered so the family had to put up with mobile stoves and the coal fire in the sitting room for five days. Things were so bad that essential supplies, bread, milk, etc., had to be dropped by helicopter in the village square. The factory too was finding it difficult to obtain supplies, so I decided to buy an ex-army four-wheel-drive lorry, which could cope with the road conditions. At last! It was now possible to get home on Friday. My abandoned family, poor things, had to suffer the cold and shortages of food, whilst I lived in the lap of luxury and was spoilt with some superb meals. Dickie was an excellent cook. Although I was not exactly flavour of the month, the family was pleased to see me. The heating oil arrived on Saturday morning and we were also able to get a lot more things at the dairy and at the mini-supermarket in the village.

The snow was still pretty thick and going for walks was quite a tricky business, as the lanes were well and truly snowed up and you could suddenly have disappeared into a ditch! The children loved it as they had never seen so much snow before and, of course, were not used to the freedom of the countryside. I made a very rough toboggan out of some odd pieces of wood and we went to one of the steep fields at the top of Christ-Cross... It was brilliant!

For several weeks we were very busy cleaning and decorating, for the previous owners had left the place in a mess. They had even ripped the toilet roll holders off the wall without bothering to undo the screws! Many of the garden plants had been taken and when the snow did finally disappear, we found that the lawn was in major need of repair. We decided that in the spring, we would get a landscape gardener to redesign the whole garden.

I would certainly not recommend moving from the town to the country in the depths of winter. Once the initial excitement of moving had worn off, Helena became very depressed. She missed her friends, and being able to pop into the city whenever she felt like it. She really felt cut off. There was only one thing to do... Buy her a car. Without warning, I took Helena down to the Speedway Garage near Silverton, on the Exeter to Tiverton road. Helena thought that we had stopped to get petrol! The garage had quite a big selection of second-hand cars and as I parked, I saw a good-looking white Mini. It had a number plate that I couldn't resist, AFJ 13. This had to be meant to be – all our girls had been born on the thirteenth. The car appeared in good shape and was very clean and tidy inside. I told Helena to get in the driving seat and I sat beside her. 'Go on and drive,' I said.

Helena was quite shocked, but she had driven a lot in Exeter, especially the Daimler, and I felt sure that her provisional licence was still current. Well, Helena had not driven for quite a while, but she was absolutely perfect. The car still had a few months left on the tax disc, so a phone call from the garage to the insurance company, a deposit with a hire purchase agreement signed, and Helena drove off home with the three girls, who had got into the Mini while I was doing the necessary formalities.

I waited deliberately for a few moments before following her home, as I was anxious not to pressurise her before she became used to the new car. When I arrived, she had already parked her car neatly in the garage and I found her sitting in the kitchen having a cup of tea with a broad grin on her face! This car was to change not only her life, but our lives also!

Helena was once again her cheerful self, although it now meant that she would have to fetch the children every afternoon from the Palace Gate School in Exeter. They came with me in the mornings, unless I was away on business.

★

Although it was a very large ranch-style bungalow, we still found that we needed another big room for Charlie to have as a study and workshop. In late summer 1963, we had the extra room built and a fantastic Spanish-style entrance with arches and a handmade real wrought-iron gate, which had as part of the design some lovely lily shaped flowers covered in real gold leaf on the outside and painted dark red on the inside. This gate was exhibited at the Chelsea Flower Show earlier in the year, by the designer who won a commendation.

Having a name like López, and after seeing the new entrance, some of the villagers decided that we were some sort of Muslims and that we had built a place of worship. The explosion in package holidays was in its early days and people were still quite ignorant of anything to do with the other side of the channel.

Within no time at all, we had a couple of ponies and a donkey. The latter came from southern Ireland, where he had been badly treated and the sores around his fetlocks were a living proof of the suffering that he must have endured. Matthew, for that was his name, probably felt sure that he had won the pools coming to live with us!

We bought a huge chalet and had it erected near the swimming pool and the landscape gardener did a magnificent job.

There was a crazy paved area between the chalet and the pool and a very grand semi-circular flight of steps leading to it, as the rest of the garden was at a lower level than the swimming pool area.

Work was going well at the factory and we were expanding at quite a pace, particularly on the plastics side. Bill sadly lost his wife after a very protracted illness. However, a few months later, he met an attractive widow whose doctor husband had died recently too. They made a perfect match and as they spent so much time together, I found myself having to run the expanding business single-handed... I loved it!

Out of the blue, we had a stroke of luck. The Crown Agents for the Colonies invited the company to quote for a special briefcase that was to take documents for the Nigeria census in the autumn of 1963. Nigeria had a census a couple of years earlier, but it was considered to be inaccurate, as the Ibo tribes at that time were quasi-nomads and many succeeded in voting more than once, simply by moving from one area to another whilst the census was taking place.

The idea was for the briefcase to be carried in bulk; 144 in one specially designed export crate, which was to weigh no more than 144 lbs in all. The crate had to be impenetrably free from insects and would be carried on a sling between two people, or preferably by one Ibo wearing a head sling, as this would prove easier when cutting their way through some of the more dense jungle.

The design submitted for us to copy was wholly unsuitable for our type of machinery and method of production. I decided to design an entirely new case constructed from RF welded and reinforced PVC. It had 13 press buttons so that the case could be folded flat and packed in bulk when not in use. I was particularly pleased with the design as I could see the potential for many applications. I decided to take out a provisional patent and I trundled off to the Crown Agents' buying offices in Millbank, London.

When I was shown into the waiting room, before seeing the junior minister, I joined a group of seven or eight others, each with a little parcel under their arm. The secretary came into the room and announced that the minister would see the representatives from out of town first. I was a little bit late getting there, but despite that, was asked to go in first. You should have seen the faces of some of the others… Especially those who had got in particularly early to get to the front of the queue.

I was greeted by a most amiable individual by the name of Mr Storey. I felt that he warmed to me and appeared most anxious to see what I had wrapped up under my arm. There was no doubt that the minister was very impressed and he thought that the price of £1.13 each was excellent. It was less than the likely estimate; in our new money that would have been £1.65. He asked me if I wouldn't mind waiting outside, or go out for lunch and come back later, as he had to interview the others and consider their submissions too. He also said that he would have to show my sample to the Nigeria office in London, as it was so different to what had been requested.

After a couple of sandwiches and a beer in a pub nearby, I returned to the minister's office but found that Mr Storey would not be back from his lunch until about 3 o'clock… So I waited. At last I was shown in and I found Mr Storey looking rather pleased with himself. He told me that the Nigeria office thought that my sample was brilliant and wanted to go ahead with an order for 240,000 briefcases to my design.

Naturally, I was over the moon, but he added after a short interval that there was no way that he could place the whole order with me, as from enquiries that he had just made, he did not think that I could make that quantity in under eight weeks. Well, I stuck to my guns and said that it would have to be all or nothing and that in any case, I had taken out a provisional patent for the product. He took a few moments before picking up the internal phone and spoke to someone who was probably more senior. The more senior voice on the other end of the phone told him that he needed his head examined, as he was going to take a chance and give the whole order to Mr López!

Bill was playing golf up in Formby, so I decided to order the extra machinery, which would prove necessary immediately, as there was no time to lose. Fortunately, all eight of the 10kw RF welders were in stock. The logistics were to prove of interest to the BBC, so they made a couple of visits with their camera crews. I had to appear on TV as part of the 9 o'clock news… This was a big order for the West Country and a valuable export one at that!

It would take too long to relate all that went on; suffice to say that we employed three shifts and worked around the clock. The first shift consisted of our normal workforce, from 7 a.m. to 5 p.m. The second shift was made up of part-time housewives from 5 p.m. to 9 p.m., and the third shift was made up of Exeter University students, who worked from 9 p.m. to 7 a.m. God only knows how they managed to get their studies done as well!

We had a visit from Mr Storey after four weeks, when we were not up to schedule due very largely to delays in getting the necessary tooling completed as well as setting up the production lines. We had to do a spot of deception... Some export crates passed by in front of him once or twice, a bit like the Generation Game! He was happy and in the end we were only three days late. The penalty charge would have been £5000 per day, but Mr Storey waived it and held up the second freighter. I had a letter of thanks from Mr Storey and also from the Nigerian Government, as my design and packing arrangements proved most successful.

There were many of my product designs that became common and in general use. Three of them were related to postal systems. One was for the blind and it became known as the talking book service; a tape recording was dropped into a zip container, which had a seal or padlock on it, with a clear PVC window on the front. This contained a name and address card, which the recipient merely turned over to return to the Institute for the Blind. The second product was very similar and was used between branches of the larger businesses, such as banks and building societies, etc. The third idea was again for use in the post by Gas Board meter readers. Instead of having to return to the office to drop off their meter readings from each area, the meter man just returned the sheets to head office by post and they in turn sent directly to his home a new batch; that way, no one had to travel. It saved not only several meter readers but also time and cost of travel.

However, I think that the alternating pressure mattresses used by hospitals and nursing homes to avoid bedsores gave me the most satisfaction. These are now in general use. It was the sheer simplicity, principle and lack of moving parts that gave me the feeling of achievement, but I owe a big debt of gratitude to Dr Ron Hadden, who had such faith in my product that he tested them for me, on real people.

★

We made many friends in Silverton and liked the village very much though the social life could prove a bit hectic at times. However, we were blessed with three good pubs, which became regular meeting places and saved on having too much home entertainment.

Charlie made a very good friend whom he met at the Lamb Inn. He was called Jack by everybody in the village and lived in a very sweet little cottage near the entrance to our drive. They became close friends; Jack had also been in the RAF, so they had something in common from the start. Charlie brought Jack home for a drink and he noticed that I was interested in oil painting. He had been one of the official RAF wartime artists. He had painted many pictures of Spitfires and Hurricanes in dogfight scenes and, to this day, several of his paintings hang in a museum in Brussels.

After a few drinks, Jack suggested that we should start a painting class in the village. As it was such a splendid idea, we went ahead with it and I helped Jack to clear his studio in order to make standing-room places for about 15 people. It was a great success from the outset. I can't quite remember the exact number of founder members, but it was in the region of 12 or so. We had our first meeting in Jack's studio/workshop, where he now had several nude paintings of young girls. They were very good and we all took great interest in them!

Only three of us had done any painting in oils before, but with Jack's help, most of us became budding artists. It was great fun, very informal. Each of us took an interest in everyone else's efforts, not only to admire, but to learn from either their success or failure! It was all most illuminating. The class expanded to 32 at one time, when wood carvers were also encouraged to join. Carving was another of Jack's talents… He was a master! With his skills and tuition, he made most members blossom. My painting technique was very different to the others', whose style was more akin to Jack's, but obviously not in the same league. I tried to change my style so that I could copy him, but without success. Jack tried my style and he too was equally unsuccessful. He told me that I should stick to my sort of 1" brush stuff as otherwise I would lose something, *'je ne sais quoi!'* I exhibited some of my pictures and a remarkable number of people wanted to buy them. But I just couldn't bear to part with them... Wasn't I silly!

We continued to spend most weekends in our workman's cottage in Abbotsham. It was so relaxing there without the telephone. The next Easter school holidays, we planned to stay at the cottage – we went in the rain and it never stopped for four days solid. We just could not stand any more, so we decided to go to Wales and it still rained, unceasingly. We stopped in Carmarthen to have a cup of tea and stretch our legs, and we found a nice café. Luckily there was a large table vacant, and the five of us sat at it and decided what to have.

The food at the other tables looked good and made us hungry. We waited and then waited some more, and we then noticed that the last two to come in were being served before us. What a cheek! We could only hear Welsh being spoken with a strong twang; we were not welcome, so walked out with a flourish. I drove like a thing possessed and we eventually got to Port Talbot. It was now getting dark and it was still raining. We tried place after place for a couple of rooms for the night, but no luck. Then I saw a commercial type of hotel as we were leaving Port Talbot. Helena was fed up with making enquiries and she was soaked. Okay. It was my turn. At last... Success! A very sexy-looking girl at the hotel reception said that she had a room and gave me a funny look and said that the room wouldn't be necessary, as I could always share with her!

At that moment, Helena walked in with the three girls, all three very close to tears. We got our two rooms – the girl at reception was too shaken to refuse. Next morning we went straight on to Barry, but the rain, if anything, was even worse. We went straight home to Silverton after that... Enough was enough.

Bill had asked me some while before that if a vacancy should occur at Tuckers Hall, or more correctly, at the Livery Company of Weavers, Fullers and Shearman, would I be interested in joining, bearing in mind that it would be expensive too. As time went on, I simply forgot about it, until one day, Bill said that the members at Tuckers Hall wanted to meet me. Would I be his guest at a banquet to be held there in a couple of weeks' time? I was delighted to accept and bought a new dinner jacket off the peg, as my old one was looking very much the worse for wear.

Bill had told me a lot about Tuckers Hall; the fact that it was only one of five in the country outside London; the fact that they still had their own hall – most had been ransacked, burnt down or confiscated by the Crown and given as grace and favour to leading families which had supported the Reformation. It appeared that most of the early guilds had religious beginnings. Many trades originated with the early monks, who were the backbone of learning and trading. The Church in those days was of course, Catholic. The reason that the London Livery Companies managed to hold on to their magnificent halls was that they were privy to the general feeling of the government being closer, and changed the status of their companies by removing all vestige of a religious connection.

It was a very grand affair. As we came in we were announced by the beadle, and Bill introduced me to the under warden, the upper warden and finally to the master. I was given some superb champagne... No less than Bollinger and as much as you could drink in half an hour! I did justice to it without letting Bill down. It appeared that you had to be a master man – own your own business and carry on that business within the confines of the city and county of Exeter – but you could not be a professional man. You had to get your hands dirty, so to speak! I was all of those things and was therefore invited to join... An enormous privilege.

I became master in 1989/90 and had to provide the under warden's feast a couple of years earlier at a cost of over £2000, as every member that was fit enough, turned up for a free evening. The feast was always held on 5th November – Guy Fawkes Night.

I did not realise at the time what a great honour it was being a member of such a select company. By the Royal Charter of 1620, the numbers were limited, however it was the limitation of dining space that kept the numbers to 24 or so! I was particularly pleased to become a member, as I was at that time the only member that had any connection with the wool and weaving trade. Helena's father had been a wool buyer for the old family business at the mill at Buckfast; previous to that, they had a seed merchants and wool business in the name of Northcott and Sons, behind the White Hart – an old coaching inn at Exeter. I regret that Helena's father died before I became the 441st Master in 1989, as he had looked forward to being the principle guest at my banquet.

I had invitations from several livery companies during my time as master. One of the most memorable was that from the Master of the Clothworkers Company, which had a magnificent banqueting hall and where the tableware was either made of silver or gold. Another impressive livery company was that of the Merchant Venturers in Bristol, who to this day, are still the official victuallers to the Royal Navy. I could mention many more, all extremely wealthy, unlike us who had to dip into our individual pockets in order to maintain our beautiful, ancient hall.

In return, I entertained the masters of these livery companies at Tuckers Hall and although we were minnows by comparison, there was no doubt that they admired the age of our incorporation, our survival and the beauty of such a rare hall outside London. I am sure that I would never have missed the experience for the whole world, and nor would Helena who, on occasions, accompanied me, for she enjoyed meeting people whom we would not have been in a position to meet otherwise. It also gave her the opportunity to dress up and look radiant as well as being very glamorous in her evening gowns.

The masters of the London livery companies were generally titled. It was traditional for the Master of the Cloth Workers to be prime candidates for the Lord Mayor of London, as had been the case of Sir Peter Gadsden CBE, AC, Master of the Cloth Workers Company, who invited me when I was master at Tuckers Hall. For my last official speech, I recounted my escape from Almería, when I was picked up from the beach at El Alquian by a British destroyer... HMS *Venezia*. They listened in absolute silence for they realised that I had only been ten years of age at the time.

Bill Bishop invited me to join Masonry after having explained to me the finer points. I became an initiate of Lopes Lodge No.

5526 in the Province of Devonshire, meeting at Exeter five times a year. Bill was a stickler for smartness and expected me to learn my lines to perfection. Woe betide me on a Monday morning at the office if I had not done things to his satisfaction at the lodge on the previous Friday night!

★

About this time, I became very interested in old cars; vintage ones, especially Rolls-Royce. However, I met this chap whose name escapes me, but he was a dealer in old Rolls and Bentley cars. I fell in love with a 1929, 20/25 Sedanca de Ville. It was in pretty good shape and the Silver Spirit was made of real silver. It was necessary to carry a spare, replacement cap so that I could lock her up in the glove compartment, when I parked in any public place.

I became a very enthusiastic member of the Rolls-Royce club and went to many functions. The one I remember as if it only happened yesterday, was the *concours d'élegance* held at Imperial Hotel in Exeter. I did a couple of things to the car and polished it to a fine glow. Then she was bloody-minded and it took me quite a few goes on the starting handle before she would go.

I made myself late, arriving at the Imperial just as the judges were about to inspect the gleaming cars lined up in rows, I was somewhat flustered, all done up as the chauffeur with the roof partly folded back. I joined the end of the line and stepped out... I fell straight on my face; my right trouser leg had somehow got caught up on the stubby brake lever next to the driver's door. Everyone burst out laughing! I got a mention. I am sure that it was as a consolation prize for the entertainment!

Helena was not altogether happy with my Rolls as it took up a lot of my time and it had to be kept in the garage, whilst the Ford Executive had to be left outside. It was inevitable for the Ford not to start on cold, damp winter mornings on the odd occasion, which meant that Helena had to jump in her Mini whilst still in her dressing gown and take the three girls to school.

Naturally, as a car of that age, even the Rolls was bound to break down on the odd occasion – and it did. Once, I was taking Helena for a jaunt; she sat like Lady Muck at the back with the glass division wound up and if we passed anyone whom she knew, she smiled and bowed just like the queen! Unfortunately, it was soon after such a flourish that the car just decided that it had had enough. There was nothing for it but for Helena to summon help. She stood by the side of the road, hitched up her skirt and a fine leg soon did the trick... Within minutes we had assistance!

The other memorable breakdown was early in the morning after a Lopes Lodge and a visit to the Musgrave Club for a couple of games of snooker. I got as far as the bottom of Silverton Hill when the clutch went. I knew how to mend it, but it needed two to do the job; one in the car by the passenger seat and the other underneath. It was 1.30 a.m. and I had no option but to walk home and leave the car in a very dangerous position.

I woke Helena and asked her to give me a tow with her Mini. Most obligingly she agreed and still half asleep, she drove down to the Rolls. Try as she may, she could not move the Rolls... Two tons was more than her poor old Mini could manage. Lots of smoke from hot tyres was the best it could do. I went back home with Helena and rang the AA. I explained my predicament, especially about the dangerous position in which I had had to leave the car.

'I am sorry, sir,' said the office manager, 'but I am afraid you have just missed the mechanic,' he said in a pretty unconvincing voice. 'What sort of car is it sir?'

'A Rolls-Royce!' I said quickly. And isn't it strange that at that point, the mechanic returned to come back for something or other, and I was told that it would be all right for him to come out. I told the office manager the exact location of the car, but was careful not to mention the age of the car in case he had made a mistake about the mechanic coming back.

Helena dropped me back to the Rolls and I think that I only waited 20 minutes before the AA man turned up. I was expecting him to get pretty annoyed when he saw the age of my car, but not a bit of it. He simply drooled over it. He turned out to be the night duty manager, as the mechanic really had left at 1 a.m. 'Do you know sir, that this is the very model on which I had my training at the Crew works,' he said.

One of us had to get under the car, but the AA man was very aware of my dinner jacket. However, I was already getting under the car at this point, and within no time at all, we had it done. I asked him to come back to the house and have a well-earned drink, to which he readily agreed and followed me in his van. I was to hear all about his life with an assortment of Rolls-Royce owners who, by and large, had been good to him. He had been with the AA only a couple of years, but was soon to retire – but not before we finished half a bottle of Scotch between us!

The old Rolls was taking a lot of my spare time and more; I was not altogether surprised to receive orders from the boss – Helena. It was going to be a case of either it goes, or I go. She told me in no uncertain terms. I was mortified! I didn't know what to do – but then I had a brainwave. Go and see Colonel Hacking! I was sure that he would help. The colonel was the managing director of Motor Macs, which was then the main Rolls-Royce dealer for the South-west.

Colonel Hacking agreed to house the car for me in an inside showroom, in case Helena should pass by and see it. This arrangement lasted for many months and during all that time, Helena was under the impression that I had sold the car, although she was curious as to why I wouldn't tell her how much I had got for it. But in the end, she decided that I had lost so much money that I didn't want to talk about it.

The arrangement with Motor Macs was to last about 18 months, but the colonel finally decided that his mechanics were spending far too much time fiddling with it! Prospective customers were also taking far too much interest in my old car, instead of the latest models!

I sold the old Rolls to Brooks, the hotel service people, at a very good price. Much to my surprise, I saw the old car several times on television over a period of six months. Brooks advertised the car with the caption 'Brooks, the Rolls-Royce of cleaners'. When the advertising campaign finished, I believe that the car was sold to an American. Each time the advert appeared on TV, Helena said on more than one occasion how much alike it was to our old car… I couldn't help but agree with her!

Helena was very proud of her Mini, which I recollect had a larger engine than usual and she enjoyed driving around like the clappers. But she was a good driver. In fact, on one afternoon she was taking her dear friend Ruby, whom she had known during the war, for a ride in the countryside. Ruby had only just finished saying what a lovely ride she was having and that she felt so safe with Helena, not like some people she could mention! Talk about tempting fate... Just at that moment, a huge Tuckers lorry came out of a field and onto the lane a few yards in front of Helena, and without warning proceeded to reverse at full speed towards her.

Helena hooted and hooted but it just kept coming on; the Mini's little 'peep peep' couldn't be heard in the cab. He was obviously trying to clean his tyres. Crunch. It all went dark inside the Mini. Miraculously, the lorry stopped trying to reverse beyond the Mini's windscreen; the differential on the rear axle of the lorry came to rest literally one inch from the glass. The front wings and the bonnet had been crushed. It was quite incredible that Ruby didn't have a heart attack there and then. They waited inside the car for what seemed an age, but they couldn't have got out anyway as the front doors were now pear-shaped!

A very nervous and feeble voice called out, 'Are you all right under there?' As the lorry driver heard the cries for help, he was mighty relieved to hear a voice. He feared that he had killed somebody. When the lorry went forward it ripped parts of the Mini's bonnet and wings. Helena and Ruby could now see daylight and soon realised that they had had a very lucky escape!

The driver admitted that it was entirely his fault and he was very nearly in tears from the relief of seeing Helena and Ruby alive and miraculously unhurt. When he had realised that he had hit something, he was too petrified to move and just sat in the cab, which explained why he did not come right away when Helena had shouted out. He explained later that he had been trying to shake off the mud from the tyres, as the police were very hot on it if lorry drivers left mud all over the main road.

At first we thought that the Mini was going to be a write-off but the insurance company decided otherwise and Helena was given a courtesy car as the Mini took six weeks to repair. From the first appearances, I thought that the repair garage had done a good job – it looked as good as new.

Helena was thrilled with her new-looking car, as all the dents and scratches had now gone. It now had new front wings, a new bonnet and radiator. I decided that if she had had a powerful twin wind horn, then the accident might not have happened, so I fitted the best that I could find – it now sounded more like a bus!

One Saturday morning, soon after Helena had got her car back, our next-door neighbour Charles, a young single chap who spent most of his time under a red sports car, asked me if Helena would mind him borrowing her car that Saturday night as his car was in bits. He had a date with a gorgeous girl... Too good to miss... You know what I mean! I said that I was sorry but he would have to ask Helena herself, as the car was worth more than my life!

Helena let him borrow her precious Mini... She always did have a soft spot for Charles. He was instructed to put her Mini back in the garage when he came home. Naturally, I expected to see the Mini in its usual place when I got up in the morning.

You can imagine my horror when I went into the garage and there was no Mini. Ah, he must have left it outside on the road... No! At that moment, Charles appeared looking sick as a dog.

'Don, you won't believe this,' he said.

'Try me,' I said in a rather terse tone of voice.

'Well, we had a lovely evening at the Nelson in Topsham and I had got my new girlfriend primed for a bit of a snog down by the river. I was full of expectancy... switched on the ignition and the car just went up in smoke!' He said this with a look of disbelief still on his face. 'We had to get out in a hurry as I couldn't find a fire extinguisher in the car. We then just stood there and watched the Mini burn,' he said.

I was devastated. How on earth was I going to tell Helena? It was obvious that Charles didn't have the stomach for it, as he was too alarmed at such a prospect, so it had to be me! Fortunately, Helena was the one who had given permission and in any case, it was Charles! After all, in her eyes he could do no wrong.

I wasn't sure whether Helena's motor policy allowed another driver, so I decided that the incident had happened to me, particularly after the recent claim. Charles had to explain in exact detail what had happened and the sequence of events. As I thought, the insurance company were taken aback to see another claim so soon after, and were not exactly going to fall over themselves to be helpful. Within a couple of days, I received a letter to say that my claim was being turned down as their inspectors had found that a new twin wind horn had been fitted by an inexperienced person. What a cheek!

I was enraged to think that they didn't believe that I was capable of doing a good job. Frankly, I thought that their findings were, to put it mildly, pathetic! I decided to employ my own motor assessor. My inspector found that the horn had been fitted correctly and that the cause of fire had been due to the loom being scorched when the new wings and bonnet were welded to the body of the car when it had been repaired. The insurance company then paid up in full, the cheque being accompanied by a rather apologetic letter!

Business went on expanding at a prodigious pace, so much so that I began to think that one of these days I could perhaps spare the time to go and see Tío Manuel. After all, Bill had been very pleased with my huge export order for the Nigerian Government via the Crown Agents. Bill even suggested that I took five or six weeks' holiday as my health had suffered somewhat after working on many occasions, literally round the clock!

Unfortunately, my thoughts of having a holiday were soon put on the back-burner as more work came in from Marks & Spencer. It was all new development and I could not leave it to others... Or, to be more precise, M&S wouldn't allow it!

Chapter Seventeen

It was a bolt from the blue when I received a very sad letter from Marisol, my boyhood sweetheart, to say that Tío Manuel was very ill, dying from cancer. My immediate reaction was to do a flying visit to Almería, but at that time I was in the thick of it, trying not only to survive, but to take a leap forward with the business, which was being stifled by lack of sufficient capital. I felt that I had to stay at the helm.

Not long after Marisol's letter, I received one from Tía Maruja, in which she told me that Tío Manuel desperately wished to see me before he died. For one reason or another, I once again failed to go, in spite of Helena's pleas. I have never forgiven myself for not having made the effort, for Tío Manuel died on 5th May 1960, uttering my name as he died. Very shortly afterwards, it became clear to me that my financial loss would prove considerable and I had no one else to blame but myself.

As my father had died so young, my Tío Manuel was made trustee for my father's share of grandfather's estate. On Manuel's death, Tía Maruja then became the trustee. However, my relationship with Tía Maruja had not been as cordial as it might have been; consequently I resolved to visit Tía Maruja as soon as possible, but again due to the pressure of work, I did not achieve my goal until August 1963.

*

My return to Spain after 37 years was bound to prove a momentous occasion, as Tía Maruja would meet Helena and my three girls for the first time. I decided to take six weeks from work and left the business in the very capable hands of my financial director.

The whole family needed a good holiday, particularly Helena, who never ceased to work until the last ounce of energy had been exhausted. We would go on the slowest P&O liner to Australia, which had Gibraltar as its first stop. Before all else, I had to check with the Spanish embassy to see if I was free to go to Spain. I had left the country in 1936 illegally, and what was more, I had not done my national service. However, I had by then become a naturalised British subject and I was assured that all had been forgiven!

We left Southampton on the one class ship, SS *Stratheden*. It was by no means large; some 25,000 to 30,000 tonnes I vaguely recall, but to Rosalyn, Rosemary and Teresa it was an absolute giant. We were all very excited, all for different reasons. I was returning home like the prodigal son and I had no idea what sort of reception I would get.

Rosalyn, being 10 years old, had a cabin on her own. The twins, now both 8, shared, and we had a cabin with twin beds, but all three rooms were next to each other. The children sat down to dinner quite early whilst we investigated one of the bars.

After dinner, the three girls went to bed. They were exhausted from all the excitement and all the running about exploring the ship before eating. Helena and I changed for the well-publicised dinner dance. It turned out to be a first-class event and we enjoyed the odd quickstep, foxtrot and old-fashioned waltz. We thought it wise to check that the children were all right... The twins were well away, but we found Rosalyn lying on her back, sobbing her heart out. She had been staring at the angle where two girders were bolted together and appeared to be sliding under several nuts and bolts over her head. It took some time to convince her that it was very necessary for the joints to be flexible, as otherwise the ship might buckle.

'Please keep an eye on the joints for me and see that they do continue to move,' I said and success... 15 minutes later, she was fast asleep.

The band played splendidly and we continued to dance into the night. We were doing a particularly good quickstep, having lost our inhibitions... The wine was delicious, so much so that I began to find that the deck appeared to be rising up and down at quite an alarming rate! Then we noticed that we were the only ones on the floor... Even the tables were sparsely occupied, but the band kept on playing. The instrumentalists were reading their music with half an eye on us. The deck then suddenly fell away and we held each other up to avoid falling. Suddenly we realised that we were in the throes of a wild storm in the middle of the Bay of Biscay!

A diplomatic exit from the floor was the only course. As we walked to the cabin, we passed several rather seedy passengers holding on to the rails along the corridor. The girls had to be in a state… Not a bit of it, they were all fast asleep.

On the fourth morning on board, we woke up early when the noise and the vibrations of the ship's engines had stopped. We had come to rest outside the harbour. The Rock of Gibraltar was shining in the early morning sun; it looked quite beautiful. A most unusual smell came wafting over from the mainland… It was a mixture of olive oil, the scent of potpourri flowers and what can only be described as benign drains! We came to associate this smell with southern Spain. Regrettably, the Costa del Sol has since been sanitised and that distinctive odour has now vanished!

We finished our last-minute packing, having done most of it the night before, and went on deck to see what was going on. A large tender had now come alongside us and as the gangway was about to be lowered, we heard a ship's officer call my name over the intercom.

'Would Señor López please go to the bridge.' I was absolutely gobsmacked. What on earth could be wrong? Was I being arrested? Had they hoodwinked me into coming to Spain under false pretences? No! They could not do that, I was still on board a British ship, I thought to myself. Hesitatingly, I climbed the stairs to the upper deck and knocked on the door to the bridge.

'Good morning, Mr López, I would like you to meet your cousin, Señor Bueno, who has come to welcome you back to Spain, on behalf of your aunt. Señor Bueno, I am sure you know, is the governor of the province of Almería.'

Señor Bueno smiled and we both said 'Encantado,' simultaneously. We passed a few pleasantries and he returned to La Línea on his own launch.

My aunt was apparently most anxious to see me again and to meet Helena and the girls. Asking Señor Bueno to welcome me was without a doubt a gesture of goodwill, to reassure me that all would be well on my return.

Luggage belonging to those getting off was loaded onto the tender first, and we followed shortly after. We were all pretty excited; the girls were ecstatic. I tried to remember what it was like when I left, 37 years ago... Nothing appeared to have changed. Helena too, had visited Gibraltar before. It had been with her parents, when on a cruise to north Africa, the Mediterranean and the Balearic Islands.

We were cleared through Customs with the minimum of fuss and found a BEA car awaiting us. The children thought that the holiday had come to an end, and that we would now drive back home immediately. They could not believe it when I told them that the holiday had only just started!

First of all, I had to go to the Rock Hotel, where I had spent several happy months in 1936. Well, it was just as good as I remembered, in fact, in had been extended and renovated. Next stop was to see the Barbary apes on the top of the Rock and to take in the magnificent view of the harbour. We then saw St Michael's Cave and part of the underground workings. Main Street didn't look very different. Next, a walk to the Catalan beach, which had lovely memories for me. We drove to the frontier where a Guardia Civil gave me a very searching look as he examined my passport, especially when he saw my name and that I was born, Almería!

We crossed in La Línea and found it less prosperous than Gibraltar, but more quaint and colourful. Unfortunately, the girls spotted a large poster advertising the *corrida* and they insisted that we had to go. However, it did not commence until late in the afternoon, so we had to hang about for a couple of hours. It was August, the heat was unbearable and we were dying of thirst. We drove about the town and came across a street vendor selling huge sundae watermelons. We simply had to have one. It only cost about 25 pesetas and we cut half into slices and had those. It was full of juice and the slices reached from ear to ear... It quenched our thirst nicely.

We arrived in the Plaza de Toros only to find that we were the first in the large car park. I decided to park in the middle in order not to get hit; what a mistake that turned out to be! We bought tickets for the shady side, *sombra*. A cool breeze made it quite pleasant, so we did not mind waiting. The excitement and anticipation kept us from being bored.

By the time the spectacle was about to start, the place had filled to capacity. We had excellent seats about three rows from the front.

The tall main doors opened suddenly with a fanfare and three flower-decked Victoria coaches entered. The first coach contained the *alcalde* and his *señora*. The *alcalde* wore black drainpipe trousers, a black bolero jacket and a black velour wide-rimmed Andalusian hat with a knotted leather chin strap. His *señora* wore a red flared flamenco ankle length dress, black platform shoes, a black lace mantilla veil draped over a high tortoiseshell comb on top of her head, a cream frilled folded square scarf over her shoulders and a beautiful cream and gold filigree *abanico* was in her right hand. The other two coaches were filled with equally colourful young men and women and civic dignitaries. The *alcalde* and his *señora* sat in the centre box with the others seated on either side.

The procession that followed was led by the brass band, the arena officials in black uniforms with red sashes around their waists and the horsemen (*picadors*), holding lances and similarly dressed, all with white breeches and white full-length socks. The horses had protective padding on each side.

Then came the three *toreros*, with red and black embroidered bolero jackets and finally entered the *matador*, dressed in a blue and heavily embroidered gold bolero jacket, a cream waistcoat, knee-length white socks, black patent buckled shoes, and, held in his right hand a black velour three-cornered hat. On his right shoulder, he carried a red cape.

The procession went around the arena and each person positioned themselves in pre-allocated places. The *matador* came to the middle, in front of the *alcalde*'s box, to whom he bowed, waved his hat with a flourish and then threw several red carnations to the prettiest of the young *señoritas*, who jumped to catch them, blowing kisses to him.

The *matador* walked to the centre of the arena and a fearsome black muira bull came charging in. The *matador* knelt in front of the beast, only rising and sidestepping as the bull came close enough to feel the heat of his breath... A chorus of 'Olé!' followed. Time after time, the *matador* made the classical passes, but then after the last pass, the bull turned quickly and more sharply, catching the *matador* in the middle of his red *faja* (sash). The *matador* went over the bull's back and landed in a heap behind the ferocious animal. There was a deadly hush. The bull turned again, with the full intention of goring to death this handsome young *matador* whilst he lay wounded on the sand.

But the *matador* managed to raise himself onto a knee and held the cape just inches away from his body, as the bull made a further charge... There was uproar... A crescendo of 'Olé!' and 'Hurray!' The *señoritas* threw their flowers, some landing by his feet. This was great stuff. But then the *picadors* moved in and placed flagged darts (*banderillas*) on the raised gristle of the bull's back, which was meant to weaken the beast.

The spectacle became more bloody by the minute as more horsemen struck the point of their lances on the same part of the bull's back. Then a ghastly sight... The bull's horn had caught one of the horses on the side, and entrails were brushed along the sandy arena floor. Rider and horse were removed behind a barrier.

The moment of truth had arrived; the *matador*, with sword hidden in the fold of the cape, knelt once more on one knee and as the bull charged with its head down, the *matador* rose, half turned and, leaning over the awesome horns, he drove his long, thin sword through the bull's shoulder and right down to its heart. The valiant muira bull died instantly.

Horsemen came and dragged the carcass away. Now I remembered why my Tío Manuel would not go to a *corrida*, even when he was the Mayor of Almería and had been expected to attend at fiesta time. As far as we were concerned, enough was enough. No way could we have faced another killing, even if the flesh was distributed to the poor and needy.

We made our way to the nearest exit only to find it barred. We tried others, but they too were similarly closed. Presumably it was done to keep out the noisy throng outside. At last, an official came to our rescue, but you could see that he was amazed at our desire to leave, from the look of incredulity on his face.

'*¿A ustedes no les gusta la corrida?*' (You don't like the bull fight?) he said, as he opened the door.

To our horror, the huge car park was full to the brim and I could see our hired car bang in the middle; there was no option but to wait and endure the constant shouts of 'Olé!'

It must have been over an hour before we were able to hit the 340 Carretera Nacional to Málaga. We had several slices of melon, but had to throw out a large chunk, as the juice was spilling over the carpet. We could only have travelled about half an hour when a Civil Guard motor patrol waved us down.

We waited in the car with the windows down whilst they parked their motorbikes. I was pretty scared as we were in the middle of nowhere, and they looked rather intimidating with their sunglasses, revolvers in their holsters, and helmets. They both walked round the car, noting that it had been hired at Gib and asked to see my family passport. One of them spoke English and wanted to know where we were going. We said Almería and they just nodded their heads and didn't seem to believe me. We had to open the boot, which was full to the gunnels with our luggage. I feared that I would be arrested as I had left the country illegally all those years ago, but suddenly, they waved us on without any explanation.

With a huge sigh of relief, we continued our journey. We went through a small village called Buenas Noches (Goodnight), not far from Estapona, when again we were stopped by a motor patrol. This really scared me, as I was certain that the previous patrol had radioed ahead of us. This lot however proved more friendly, but asked us to get out of the car. They searched thoroughly and then told us to carry on. I now felt happier, so as we left, I ventured to ask them what they were looking for and was told cigarettes, drugs and cameras. They also said that it was doubtful if any other patrol would stop us.

We were off yet again, but it was getting a bit late and we felt that it would be sensible to stop at Estapona while it was still light out. We saw very few hotels in the town, but we found one on the beach, which looked very new. We booked two ground floor rooms with patio doors onto the sandy beach. It was heaven and we were determined to relax and enjoy our first dinner in Spain after our first swim in the Med.

Reception found it hard to believe that I had been born in Almería in view of my English looks and English accent. They also didn't think that the road went much beyond Málaga; what rubbish – as a small boy I had travelled with my family on more than one occasion from Almería to Gibraltar!

Next morning we had a big buffet breakfast and made an early start, as I gathered that the road deteriorated as you went east. The next few kilometres to San Pedro de Alcántara were quite uneventful. We decided not to stop and pressed on to Marbella where we intended to stay the night. We found a new hotel by the port, which looked very attractive. The road so far had been reasonable, although the tarmac surface tended to melt under the intense heat of the sun. Any sudden braking was somewhat hazardous, as the stone blocks at the edge of the road were often broken or non-existent... A bad skid could send the car careering down a precipitous rock face, which ended up at the bottom of a ravine, some 1000 to 1500 feet below. We saw more than one wreck of a car that had plunged over the edge; the plight of those who might have been in those vehicles did not bear thinking about.

Tourism was very much in its infancy; consequently, new hotels were few and far between. Many of the early hotels left a lot to be desired. We were lucky with the Bellamar as it was excellent. We booked two rooms with balconies overlooking the harbour. We spent the afternoon walking around the very quaint old town. It was more like an overgrown village than a town and it seemed as if it had been caught in a time capsule. The narrow roads radiated from a square in the centre of the *pueblo*. On one side stood a large, circular fountain around which several old women were having animated conversations, all dressed in black, their heads covered by shawls and resting their elbows on large water pitchers.

As we passed, the locals gawped at us as if we were visitors from outer space! It gave us an eerie feeling, but we did not fear them. On the contrary, I wanted to talk to them... After all, I could speak Spanish and that made them even more intrigued.

Dinner started early by Spanish standards. It was good... the girls enjoyed their first Spanish tortillas with *jamón Serrano*...

They also enjoyed a glass of strong local rasping red wine. We thought that the girls should be introduced to wine, and in any case, we did not wish to risk tap water, although I am sure that it would have been good, for it would have come straight from the Sierra Nevada mountains close behind us.

After dinner we sat on the balcony and watched the locals *pasear*. This consists of the family walking up and down the main street, or *paseo* in order to show off their offspring, the youngest males blowing *piropos* at the *señoritas* in their very best finery, as they walked side by side with their ever-watchful mothers.

We were happy to be entertained by such a colourful spectacle, when suddenly we noticed a well-dressed elderly man walking in between a group of passers-by, holding a dog lead, but at the end of which was a small pig, not a dog! Perhaps he too was looking for a likely mate.

Although we had not travelled far, we felt that it would be wise to have an early night for I had every intention of getting as far as Málaga the next day.

Once again, we had a large buffet breakfast, avoiding a stop for lunch, ice-old drinks being the order of the day. Very soon, we were on the main coastal road, N340, which we later learnt was otherwise known as the 'road of death'… How right they were!

The road proved to be very similar to what we had encountered so far, and we did stop after all, not only for quick relief by the roadside, but we also found small villages on the way which we found of considerable interest. We also came across several goat herds with a variety of bells hanging from their necks, being driven along by shepherds who had deep, sun-baked faces and who were in no hurry to make way for us… We had no alternative but to wait for a convenient place to pass as we listened to the very melodious bells.

With one thing or another, all our good intentions went out of the window and we only got as far as Fuengirola, for we could not resist stopping several times to pick the fruit of the prickly pears, known as *chumbos*, which I remember liking as a small boy. They were difficult to peel, as we had no gloves to protect ourselves from the hundreds of dreadful spines.

We found Fuengirola a most attractive place and decided to stay the night. We found a hotel close to the beach and although the road ran along the coast, the heat had been so intense in the midday sun that we found the azure sea too inviting to ignore. The sea was gloriously warm, so much so that we did not realise how long we had been swimming! Our knuckles had become white and our fingertips wrinkled like prunes.

The food at the hotel was very similar to the previous hotel, so we had no problem with it, particularly as we were starving. The wine too was good, and what with all the excitement as well, we were quite ready to go to bed early that night.

Another early start the next morning after a substantial breakfast and this time we did get as far as Málaga, after having stopped and looked around Torremolinos. I am sure I caught a cold there, for we went into a supermarket that was shaped like a boat and it was air conditioned; normally for that we would have been grateful, but we found the contrast to the sweltering heat outside just too much.

Málaga, unlike the other towns on the way, was a big city with a large working port, which handled the major wine exports for the region. It was just after two in the afternoon and I felt that it would be an appropriate time to visit my old school, which I had left in a hurry 36 years earlier, leaving my brother Luis without even having the chance to say goodbye. At the time, I had no idea that the communists had murdered most of the masters, most of whom were monks, soon after I had left, so I was rather expecting to see some familiar faces when we called. It was rather silly when I came to think about it. Some were quite old already, 36 years before.

Anyway, much to our delight, the janitor at the gatehouse said that he was sure that the headmaster would want to see us. We were shown into the headmaster's office, where we waited for a short while before a most genial, tall monk in a black cassock approached me with an outstretched hand to give us a most cordial welcome. I could not recognise him, although he was getting on in years. Much to my surprise, he recognised the name and added that I had an older brother called Luis. We were offered refreshments, but all we wanted was a glass of iced orange juice. It turned out during our conversation that he had been a novice and was able to escape when he saw the massacre taking place, as he hid behind one of the many pillars surrounding the inner courtyard.

We were given a tour of inspection, which in spite of the many years that had elapsed, confirmed the vivid memories I had always had of the school. I had forgotten very little, although I must admit that I did not recollect the size of the school... It now seemed so much bigger! This was most unusual, as childhood memories tend to exaggerate the size of things in the past.

I had often described the inside of the chapel to Helena, as it was so beautiful. Again, the chapel appeared larger than I had remembered it. I had of course forgotten that it also served as the local parish church and as part of the seminary of the Salesiano order. This order was worldwide and was founded by Don Juan Bosco, originally for the purpose of teaching the poorest children, not unlike many of the public schools in the UK. The first Salesiano school built in Spain was in Sevilla.

I believe that the school chapel is based on the Basilica de María in Turín. As we entered it, we could hear the strains of the organ being played as background music. I noticed that there were several altars, at some of which Mass was being celebrated. The Salesiano monks had to say Mass every day of their lives, unless they were prevented from doing so. For this reason, it was quite possible for two or more Masses to be heard at any one time... Unlike the Christian brothers, who were not ordained as priests, and hence Prior Park had its own chaplain in order to celebrate Mass everyday. Students were obliged to attend, unless you were not Catholic, or you had to have a jolly good excuse!

We finally ended up by the organ as the monk who had been playing had finished. The headmaster introduced me and left to return to his office. The organist was not only very old, but also appeared to be totally blind. I spoke to him in Spanish and introduced Helena and the three girls.

'Sí... me recuerdo; es Emilio, el hermano menor de Luis López Echeverría... un vasco.'

Well, you could have knocked me down with a feather... it was unbelievable. We spoke for a while and he wanted to know if

I was a good Catholic and that I was bringing up the children in the only true faith. I didn't have the heart to tell him the truth so I just smiled.

This monk had been the choirmaster during my time there and had known many of the boys. He was called José Cadenas, and was still able to play the organ, despite being totally blind.

During our conversations, he described to us how he had survived the massacre. He had been lucky when the communists burst into the school and machine-gunned down all the monks. They left him for dead, but he had not been mortally wounded. The murderous gang then set the chapel alight and desecrated everything before them. I was to learn from my brother that he had been one of the lucky ones. The anarchists shot 14 of the masters, including, Estéban Enarejo, subject – tailoring; Don Antonio Martínez, chief of engineering; Don Tomás Alonso, head of printing; Don Antonio Dancorvo, head prefect; Don Francisco Gómez, secretary; Don Antonio Arena, priest.

Their bodies were dragged out and left at the side of the road, to show passers-by what fate could await them if they were connected in any way to the church. Luis, terrified at seeing these atrocities, did not wait for the return of the murderous gang. He ran down to the port where he saw a ship lying in the harbour, called the *Montero*. It took him to Cartagena. He then walked to Almería, getting the occasional lift on the way.

We then excused ourselves as the music was once again required. We watched from a distance to see how he was able to play without his sight... His head inclined as if he were looking up to heaven.

It was now four thirty so we said our goodbyes and left. Our next stop should have been Almuñecar, but the road was distinctly worse and our progress became abysmal. The road surface was terrible; there were potholes everywhere and we hit the exhaust of the car several times. At last, we arrived in Nerja; it was now getting dark and we had no idea if we could find a hotel at that time of the evening.

We followed the general direction of family walkers and ended up in a square with trees on all sides, under which there were tables and chairs. This was obviously the centre of Nerja. On a dais more or less in the middle stood a policeman in a white uniform and wearing a white, pith sun helmet, directing the traffic, such as it was – mostly horses and carts stacked high with alfalfa animal feed. The square was bustling with people and we stopped by the policeman, whom we noticed appeared to have plenty of time to talk to passers-by. We asked him if he knew where we could find a hotel for the night.

Our jovial policeman didn't reply immediately. He gave us a look of disbelief as he bent down to see the car. He stroked his chin and said that he didn't know of one, but suggested that we could try a shop that sold piece goods not so far from the square, as he knew that they took in paying guests. He could see that I was finding it difficult to understand his directions and suddenly called out, 'Paco, iven aquí!'

This small boy, who looked more like a ragamuffin, had been playing with a friend near the drinking font, and came over to us. Paco obviously knew where to go and the policeman asked me if I minded if the boy rode in the car with us. It was a very short distance before Paco indicated for us to stop. We were outside a double-fronted shop, which appeared closed and barred. Paco banged on the dark oak door and presently we heard bolts and bars being drawn back. A woman of about 50 years of age appeared; she had a kind face and gave us a warm welcome. Much to her surprise, I spoke to her in Spanish. Yes, she would be happy to have us. Paco gave a long low whistle as I gave him a 100-peseta note… He scarpered as fast as he could go.

We were shown the *cuarto de matrimonio* with a king-size bed. That was to be ours. The three girls were put together in another equally large room with a double bed and a single bed. *La señora* said that it was rather late to get a big meal, but would we be happy to have 'una tortilla grande con patatas'? I couldn't think of anything nicer. In no time at all, we were shown to their small dining room and were served with the most delicious omelette and Spanish salad, which included wedges of beefy tomato, olives and crispy lettuce.

Helena and I were pretty tired; Helena particularly, as she had driven all the way from the passenger seat! The children would have liked to have stayed up, but orders from she that must be obeyed saw them off to bed. We hadn't slept in a double bed since our honeymoon, but there was room enough and we were about to drop off when Rosalyn burst in to tell us that there were cowboys down below next to the shop. We had to look out of the balcony only to find that some *rancheros* had come down from the sierras to drink at a tavern on the other side of the road. They tied their horses to a wooden rail in front of the outdoor cinema whilst drinking and occasionally letting off the pistols they were wearing over their *chaparajos* (chaps). Rosalyn was particularly excited, as one of these cowboys was all dressed in black with a shallow *almeriense* velour hat with a chin strap, just like the cowboy in the TV series, *Laramy*. Next morning, we decided to stay there another night, but we would eat out at a restaurant we had seen on the way in called the Portofino which was built over the sea.

Breakfast was not quite in the hotel class, but nevertheless it was more than adequate. We had *bunuelos* (sort of doughnut), boiled eggs, *jamón Serrano*, toast and lashings of strong coffee. The girls were happy with a glass of goat's milk, still warm, as the herd had called just before breakfast.

We had a mooch around the town and got carried away with some very pretty fans (*abanicos*); they could be very useful in the intense midday sun when the air became stifling. Having taken longer than I had calculated, we were getting short of pesetas, but no bank in sight. We decided to sit at one of the cafés in the square and have a thick chocolate drink with *churros*, which are sort of batter sticks covered with sugar. These could be dunked into the thick chocolate... Most delicious, and it brought back some very happy childhood memories.

I felt sure that if anyone would know how to get some traveller's cheques exchanged, it would be the *camarero* who served us. We were in luck, although I found it difficult to believe – he said that I had to go to someone's private house for the transaction! I began to wonder if it might be a con! I should not have doubted the waiter, for sure enough, when we knocked on the door of the house that the waiter had indicated, a rather serious-looking man opened the door. We were ushered into his front room where an enormous Chubb safe stood at the other end of the room. He was very business-like and gave me a better rate of exchange than I had expected.

We found the shoreline very beautiful and it had a small beach tucked into a cove near the Portofino restaurant. It was ideal and safe looking, but we hadn't brought our bathing costumes with us, so we contented ourselves with some glorious sunbathing instead. We returned to our rooms quite late in the evening. A quick wash, a change of clothes and we strolled down to the Portofino.

We were seated at a table on the balcony overlooking the sea, and enjoyed a couple of drinks before ordering dinner. It was getting dark by the time we had decided what to have, but it had to be fish. We could now see that a storm was brewing far out to sea; great flashes of lightening, followed by distant sounds of thunder.

The sea near to the balcony appeared like a mirror and reflected the restaurant lights as pulsating beams. It was very romantic and we made up our minds that we should come again next time we visited Spain. We lingered over a coffee and *Fundador* brandy and wished that we could stay longer, but we had a long way ahead and the road, we suspected, was going to get a lot worse.

We made our way in the morning a bit later than intended. The road did become progressively worse, so much so that Helena and the children were getting somewhat agitated, as not only were the potholes getting bigger and virtually joining up, but the road in places had fallen away into the ravine about 700 or 800 feet below. We stopped in a village called Las Palomas and a small boy pressed his nose against the front passenger window; he was holding a spear with a large squid transfixed. I think he was trying to sell it to us. Several people came out to look. A car was quite a novelty. I asked if the road got better.

I got a mixed reply. Some said the road *era muy malo* (was really bad) and others said that *no hay camino, señor* (there is no way through, sir). It really didn't matter who was right; neither was good. We went on; truly I did not think that it could get worse, but it did. Our wheels were desperately near to the edge; sometimes a tyre actually straddled a gap on the road, and the twists and curves of the road were hardly believable. Then as we went around another hairpin bend we came to a stop... Some workmen were clearing a landslide; they were prisoners. A guard stood over them with a rifle slung over his shoulders. I got out of the car with some difficulty, with only a slight margin to edge past the car.

The guard was surprised to see us. As I stood talking to him, I noticed that there was a break of about 10 to 12 inches wide across the whole width of the road. The girls had seen it too! In response to my question, the guard said that he didn't remember seeing a car on that stretch of road since the war had ended, but if we had got so far, well, the rest of the road to Almería was about the same, but we would have to wait until they cleared some of the boulders and rubble.

Luckily Helena and the girls at this time did not understand a word of Spanish.

'So what did he say about the condition of the road ahead?' enquired Helena.

'Much better, he said,' I replied.

Without a further word I got into first gear and drove at speed over the gap in the road and over a heap of rubble, tilting the car dangerously towards the edge... Screams all round, but we made it. What a relief... My heart had been in my mouth.

Strangely enough, the road further on was no worse, or maybe we had become used to it. Whatever the case, we made good headway to Almuñecar. We were all desperate to go to the loo. The first place we got to was Hotel Sexi. We trooped in and asked if we could use their toilet, as we had not seen a public one on the way. The girl at reception was delighted to show us *el retrete*.

The three girls went first and immediately came out again.

'Mum, it's only a hole in the floor!' all three said at once. We had a look and sure enough, they were right... It was complete with a brush to clear up for those whose aim left a lot to be desired. No time to stand on ceremony! We did not have time to look around, as we had to press on. We wanted to get to Adra and the road was certainly better, but the tarmac was melting and consequently, the hairpin bends really did have to be negotiated with care.

We just began to relax when the road between La Rabita and Adra became undriveable at anything over 20 miles an hour. As a result, we got to Adra very late and had to stop at some rough transport accommodation at the roadside. There were several *camiones* parked outside with girls' names across the tops of their cabs, such as Carmen, Anita or Pepita and so on.

All conversation stopped as we entered. Mostly men were standing by a bar drinking *cerveza* out of the bottle. The owner came over and tried to make us feel welcome and said that yes, we could have a room for all five of us. We sorted out who was sleeping where and each went to the one and only toilet on our floor. No shocks. It was a little better than Hotel Sexi. We sat in a corner away from the bar and ordered a couple of San Miguels and three Fantas. It would seem that we were too late for any hot food, so we had to have an assortment of cold meats; *jamón de York*, *salchichón*, *chorizo*, *queso*, *bocadillos* and *mantequilla*. Helena, I'm afraid did not have very much, not eating meat... Never mind. I would make it up to her when we got to the Eritaña restaurant, which was just before you entered Almería, if my memory served me right.

I paid the bill before going to bed, so we were able to leave early in the morning, and hoped to stop for lunch at the Eritaña, leaving ourselves plenty of time to arrive at Tía Maruja's by about three in the afternoon. The surface of the road was somewhat better now, but blind bends made driving a nerve-wracking experience.

We were regularly confronted by *camiones* coming out of a blind bend on the wrong side of the road; they were too big to negotiate the narrow road and it proved a nightmare. At last, just as I remembered, there was the Eritaña on the right, next to the sparkling blue Mediterranean; a little smaller than I had remembered. The speciality at the Eritaña had always been *cigalas*, *langostinos* and *gambas* – large crayfish and Mediterranean prawns. We sat by an open window overlooking the sea and ordered *una plata grande de gambas y cigalas*. Within minutes, an enormous oval plate arrived with the shellfish piled up high, a heap of bread rolls, a real *sangría* with a brandy base and bowls of lemon water to wash our fingers. It was a magnificent feast!

We left just after three and about two miles on, we saw Almería in the near distance; it was a shimmering with heat. We could see the port that my grandfather had built… I was home after 37 years!

Surprisingly, I found my way through Almería to Chocillas without difficulty, even though it is a big city. I went up the *paseo* as I had done with my father and Tío Manuel. Sadly, I would not see either of them again, but my heart was bursting with joy at being home again… I had dreamt of my homecoming so many times. Helena became caught up in my excitement and was happy for me. In my moment of utter jubilation, I turned left at the roundabout at the top of the *paseo* instead of right… For a moment I had forgotten we were supposed to drive on the right-hand side of the road.

A policeman who had been standing on a dais in the middle of the roundabout, with an umbrella over his head, came over and signalled for me to stop. I was sorely tempted not to stop, but Helena very wisely insisted that I did and what's more, she begged me not to speak to the policeman in Spanish. How right she proved to be! Our municipal copper walked around the car, saw that it was GBZ plates and proceeded to tell me what he thought of me, my parents and the English in general. Although my temperature had risen to smouldering point with his disparaging remarks, Helena was able to hold me down with a firm hand… not appearing to have had any effect on me, he gave up and waved me on my way with a most contemptuous gesture of his hand.

Very soon after, we were at the Barrio Alto, where the family had the big house, El Depósito. Then we passed the dry river Andarax followed by the Convento de Las Hermanas de Pobre… My old nursery school. Then finally, *el manicomio*; the loony bin. Now we had to turn left by the *fuente* at Cuatro Caminos. We drove up to Chocillas over a newly made road. Previously, it had been a rough track carved out of volcanic rock.

My excitement was beyond bounds as we turned right into a narrow road with high walls. If we had turned left we would have arrived at La Torre, which had been owned by my grandfather, but if we had gone straight on, we would have arrived at La Pipa, where I had lived in the earlier part of my life.

We passed Villa Anita on the right with large wrought-iron gates, where Tío Manuel had brought me after rescuing me from the *hospicio*; a children's home. Then I turned immediately left through a large entrance, the gates of which had been left open for us by Juan, the gardener. He lived at the lodge by the entrance to Villa López. Unfortunately, in my state of heightened excitement, I rashly turned the steering wheel too sharply... Crunch! The left-hand side of the new hire car hit the unforgiving gatepost at midships. I was gutted to think that we had come through the most dangerous and tortuous terrain without so much as a scratch and then this! I am afraid that the triumphant entrance I had visualised was somewhat dented! After examining the damage peremptorily, I received a fond welcome from Juan and his wife María; both had been in service with my aunt as youngsters and could remember me as a small boy. They were crying with joy and were delighted to meet Helena and the children. They told us that my aunt had not been sleeping in anticipation of my homecoming.

We went up the half-mile drive, lined with almond trees on either side. We could now see Tía Maruja waiting at the edge of the patio with hands clasped in front of her and looking down the drive watching our arrival. She looked diminutive, much shorter than I had remembered.

Tía Maruja, I recollected, had never been terribly demonstrative, but on this occasion she became emotional and actually cried! Needless to say, I was overcome too by the significance of the moment... I had come home! I said nothing but '*Hola, Tía,*' and held my embrace with her for a long time. I introduced Helena, who looked positively radiant and the three girls, all looking gorgeous. Big hugs and smiles all round, plus the various forms of gesticulations... Tía could not speak a word of English, but somehow or other, we all understood what the other was trying to say. Only now and then did I have to translate the more tricky phrases for which they appeared not to have a suitable gesticulation.

We reached a sort of lull in the conversation, when dear old Pura came out of the house carrying a tray. We were going to have tea on the patio by the French windows, away from the sun. I hardly recognised Pura, she had aged so. When I left in 1936, she was a lovely looking girl. She called me *señorito* with a lot of warmth in her voice. She never did say much though. I gave her a big squeeze, she had been such a good friend, never telling Tía what I got up to!

After tea, the girls wanted to freshen up and asked to have a bath. Helena wanted to look around the house and recognised a great deal from the many descriptions I had related to her. I think she was quite impressed. Tía took her to our quarters, which were through the oak-panelled dining room where several oil paintings of the family covered the walls. Hanging over the dining table was the most magnificent hand-carved chandelier depicting six dragon heads, which had been converted from candles to electric fittings. The study door, like the others, was oak-panelled with a hand-carved ornamental feature over. You then walked through the study and library, which had a desk and matching chairs, with similar features to the chandelier in the dining room; all had come from a Spanish galleon. By contrast, Tía's suite of rooms was very modern and had a walk-in bath in her ensuite bathroom, a sitting room and separate kitchen. These together made her living quarters totally independent from the rest of the house.

When Helena and Tía returned from their tour of inspection, Helena decided that she too would have a bath before dinner. Later, just as we all sat down in the patio enjoying a drink, Pura came over and whispered in Tía's ear. Obviously a bit of a problem had arisen, from the look of alarm on Tía's face. It appeared that the girls had emptied the tank on the roof heated by the sun!

Unfortunately, I had forgotten to warn the girls that the water supply in the Almería area was quite critical. We had seen several reminders of the desperate water situation as we drove along the coast road near Almería; white daubs of paint were on the walls of farm buildings everywhere, with the words *'Franco, Franco... Agua Agua'*. It would now be the next morning before water was switched on again for about three or four hours when you were able to fill up all the tanks and irrigate the gardens.

Despite the water fiasco, Pura put on a wonderful five-course dinner, which we ate on the big white marble oval table on the patio, served by María, Juan's wife. It was a perfect balmy evening with the sky studded with twinkling stars. Over dinner, Tía told us that General Franco, now called El Caudillo, had sent her a wild boar which he had shot in a hunting expedition. She had no option but to have it buried in the garden!

When we went to bed, Tía told us that Helena would be sleeping in the same bed as Peter O'Toole had used and Rosalyn would have the same one that Omar Shariff had slept in during the making of Laurence of Arabia by Horizon Films at El Alquian. She had let part of the house to the film-makers. When they left, Tía had to hire a skip to clear the pantry of empty whisky bottles left behind by Peter O'Toole, who incidentally, had taken a shine to Tía and still kept in touch with her at that point.

Horizon Films had to inspect the house after the film was completed as the two stars had done considerable damage, no doubt when they had their wild parties with some of the entourage, which had included a number of rather scantily dressed mademoiselles. The window cleaner had certainly increased the frequency of his visits! Peter, in one of his wild moments, prised an enamel figure from the face of Tío Manuel's treasured Louis XVI mantelpiece clock which had a protective glass dome.

The next morning, we went to see Tía's sister, Soledad, and her husband, who had never fully recovered his health from the privations and ill treatment he received from the Rojos during the Civil War. We had lunch with them and Tía stayed behind when we went to see my sister, Anita, whom I hadn't seen for 37 years. It was another emotional experience. I still had a picture on my mind of a very pretty young girl in a school uniform, yet now I saw a grown-up woman with a husband and four children. Anita was thrilled to see me and my family. Helena took to her immediately and somehow or other, they were able to have long conversations.

The sitting room, being on the first floor, had a balcony overlooking the narrow street below, which was just off the *paseo*. Sitting on the balcony, was my Aunt Luisa, who was now living with Anita, and although she was now an old girl, she still looked regal and attractive. She never married, although she had been head over heels in love with Gerónomo, but as the family had considered him unsuitable for her class, nothing ever came of their feelings for each other. It was lovely to talk to her and hear so many stories about the family as I had been severed from it for so long.

Anita's husband, Angel, was a professor of physics and mathematics; a rather reserved intellectual type, but you could see that he was devoted to my sister. Late in the evening, Anita's children and grandchildren arrived. It was obviously to be a feast and a get-together in our honour! Our three girls had no idea that they had so many cousins, uncles and aunts... Mind you, nor did I!

Champagne and wine began to flow and a general *jaleo* ensued. In an hour or so, food began to arrive from the kitchen until the table groaned with a multitude of dishes, from *langostinos*, Spanish omelettes, *salchichón*, *enpanadillas de atún*, *jamón Serrano*, cold chicken, asparagus dipped in butter, *chuletas de cordero* as well as *ensaladas de lechuga, cebolla española y tomates grandes en vinagre y aceite de olivo vírgen*... And if that wasn't enough, a great *paella de mariscos* was placed on a table nearby. I am sure that there was much more, but my eyes were overwhelmed!

Finally, bowls of fruit and very sweet gooey cakes were put on the table, together with slices of two kinds of melon as well as numerous cheeses and an excellent vintage rioja to wash it all down with. Our little girls noticed that their cousins were pretty free with the wine, so we found it difficult to restrain them. Fortunately, none was the worse for ware by the time we made our way home.

Naturally, in the excitement, combined with liberal helpings of an assortment of alcohol, everyone lost any shyness or inhibitions they might have had. It was in this sort of atmosphere that my sister informed me that I was older than I said. Not believing her, she promised me that she would get a copy of my birth certificate. This she did, when we came to Almería a couple of years later. It appeared that when I came to England, my guardian to be had told the immigration people at Tilberty that I was born on 5th June because he remembered that was when the family always celebrated so naturally he thought my birthday was on that day. In fact, it was my Saint's Day! This is quite normal in Spain and other predominantly Catholic countries. It only took a moment and I became over four months older – my real birthday being 25th January... This became a real body blow!

When we arrived back at Villa López, we found that Tía had already gone to bed and María had been waiting for our return before going home down to the lodge.

Next morning over breakfast, Tía informed us that the village people would be coming to celebrate my homecoming and to make sure that we would be available. Nevertheless, Pura was packing a picnic for us and Tía expected us to go to the beach for the day. We decided that we would go to Agua Dulce where, as a small boy, I would go to the company chalet with the family or with Charlie. The chalet had been built partially over the sea, which meant that you could hear the ripple of the sea under your feet. It had a veranda on three sides and was capable of sleeping eight people.

Helena and the children couldn't get over the beauty of Agua Dulce, where the sea was crystal clear and you could see a variety of colourful fish swimming around near you, as if without fear. The chalet was about a mile beyond Salinas, where the sea was let into two vast enclosures, and the water was allowed to evaporate under the intense heat of the sun, which sometimes climbed to over 113 degrees Fahrenheit in the shade. The salt left behind was then scooped up by an endless queue of *camiones* and taken away.

Pura had made two gorgeous Spanish omelettes which were made with more eggs than usual and contained chopped potato, *jamón de York* and onion... They were still hot when we each had a couple of thick slices and it was delicious! The beach at El Alquian was good, but Agua Dulce was something else. The sea was so calm – it did not have the powerful rollers of El Alquian, which kept coming at you with a force that could knock you over. Agua Dulce lies in a sheltered bay and because of the very small movement of the sea, the water becomes warm and crystal clear. There is no sand, only fine colourful ground coral.

We were reluctant to leave, but we were mindful of Tía's wishes and drove back to Almería and on to Chocillas. This time we were careful with the water; after all, we did bathe in the sea. Anxious not to be late, we were downstairs in no time at all in our best bib and tucker, when Tía arrived in a black lacy outfit looking like she owned the place!

Pura bought a tray of glasses and a large *jarro* of iced *sangría*. She had added a *copa of Fundador*... What a difference it made to the *sangría*. We were about to have a top-up when we saw coming up the drive a couple of dozen people dressed in all their finery.

Tía walked down towards them and led the group back to the patio, where we were standing in a semicircle. First was Carlos, the eldest son of the grocer. I had known him well. 'Un fuerte abrazo,' he said as he put his arm over my shoulder. He was followed by Juanito. Yes, of course I could remember him. Then came my very best friend whom I was officially not allowed to play with as a small boy... He had always been his own person, totally independent and always in scrapes with the Guardia Civil; a most exciting person. He was called El Pescadero by those who considered him a likeable rogue and a *ladrón* (thief), and by others who had nothing but contempt for him, although he was, by trade, a successful fisherman!

The introductions went on for some time as each one wished to say something in order for me to remember them by. They were more than grown up, they were mature, and some brought with them their children too. The patio was large and accommodated everyone with ease, even though more people kept coming. I think the whole village was there by the end of the night.

Several had brought small casks of wine and were busy dispensing drink to everyone and, within a few moments, dancing started as we heard the strains of music from more than one guitar. The rhythm sounded familiar... It was a *pasadoble*!

Tía had helped Pura to prepare a buffet supper. This was placed on the marble table and was quite a spread. Light was now fading and the strong outside electric lights came on. Perhaps this was the cue for a most elegant couple, dressed in traditional Andalusian attire, to take the floor and dance a lively flamenco with a superb guitarist accompanying them, who was no mean amateur. He knew his onions, you might say!

The audience accompanied the dancers with rhythmic clapping of the hands and some with castanets. This had turned out to be a really wonderful evening. I had no idea that the village folk felt strongly enough about me and the family to put on such a display of warm affection. I was quite touched and I knew that it had made a great impression on Helena and also, the three girls. It was something that they would never forget.

It must have been after 2 o'clock in the morning when the last revellers left us with our incredible memories. This night had brought home to me the very Spanish atmosphere that I had forgotten for so long. I didn't know whether we could cope with any more excitement before we went home, but for sure, there would be more; we still had to go to Barcelona to see my brother and his family. That would be a whole event on its own! But for now it was bedtime for everybody, even though the girls would have liked more.

Next morning, Tía informed us that her very best friend, Rosario, would be coming to tea and to play canasta. As we had very little time before Rosario came, I decided to have a walk about with Helena and the children. I took them to see the huge *balsa* beyond the kitchen garden where I used to swim, although I wasn't allowed to, as it was very deep and was our main reservoir for the irrigation of the numerous plantations. The frogs were there sunning themselves on the edge just as they used to. They plopped into the water as we approached.

We went out of the back entrance and walked to La Pipa, which was only a couple of hundred yards away. It had that beautiful arched entrance with the wrought-iron lanterns on either side. We called on my cousin, also named Don Emilio, but I think that we had called at an inconvenient time, as he kept us talking in the hall. I had hoped to show Helena the inside of the house as I had lived there for quite a while in my youth. He suggested that we came for drinks one evening, but there was more to his apparent coolness. He had inherited the *hacienda* when I was done out of my inheritance, on my grandmother's signing away of much of the property which was due to my father. Our unheralded visit was a bit of a shock and embarrassment, as I discovered after talking to Tía. Apparently my cousin had always felt quite guilty about the inheritance.

As we still had time left, we went on walking to La Torre where my grandfather had lived for many years. When my grandmother died, my grandfather sold it to Don José Romero Balmas; a very wealthy wine exporter and father to Marisol, my first girlfriend. Unfortunately, Doña Josefina and Marisol were not at home, but the housekeeper was more than happy to show us around. I wanted Helena to see the pink bathroom and the alabaster bath in the shape of an oyster shell. My grandfather had had the shell specially carved by his stonemasons at the family marble quarry at Olula up in the Sierra de los Fibrales. He told my grandmother that it was fit for a queen. She was the Marquesa de Patrullo, but not quite the queen!

I had forgotten how large La Torre was, and what a marvellous panoramic view one had from the top of the tower, viewing the 2000 or so acres of the rich, flat, fertile land of the home farm. I would have liked to walk around the magnificent gardens, but time was getting on and lunch was about to be ready.

Every day we had something different for lunch and dinner. Today we had iced gazpacho soup, followed by a *cocido* – a sort of meat stew, with potato, *guisantes* and lentils. We had stuffed green, red and yellow peppers with *patatas fritas* for the main course and dessert generally consisted of *flan* (crème caramel) and fruit.

Tía went to her quarters for a siesta and we sat in the shade for a couple of hours. Rosario arrived in her chauffeur-driven car. Her husband, who was the owner of the McAndrew shipping line, was in London. Rosario was quite a formidable-looking woman who I could imagine would not tolerate any nonsense; she put the fear of God in me!

During tea, before Tía and Rosario began their game of canasta, Tía, forgetting for a moment that I spoke Spanish, asked Rosario if Helena spoke good English and what sort of class did she come from? Well, I was horrified! Horrified to find Tía such a snob. 'Calm yourself, Maruja,' replied Rosario, 'she comes from the best.'

Every day there was something to do or someone to visit. There was never a dull moment. Today, we were to go and see my Aunt Antonia, mother of Julia Casinello, an internationally known architect. His brother Andrés, was a general and chief of staff of the Guardia Civil and their sister, Mercedes, head of the *Cruz Roja* – the Spanish Red Cross. Antonia's husband, my uncle, was murdered by the Communists during the Civil War and a square in the centre of Almería was named after him, in his memory.

Both cousins had charming wives and our girls got on well with their children. That evening, the smallest of their girls was appearing on the main stage of the theatre in Almería, where she performed a flamenco dance, dressed in a beautiful red and white polkadot traditional long dress.

We were about to take our leave when Mercedes informed me that there was to be a Red Cross banquet the following night at the Club de Mar and that she would like us to come and would like for me to say a few words in Spanish after her main speech. I can only describe my immediate reaction as nothing short of utter panic! Sensing that her request might bring about an embolism, she put her arms around me and endeavoured to give me good cause for the proposed short speech.

'Both branches of the family are synonymous with the history of Almería,' she said. 'The knowledge of your return from England will be of great interest to everyone.'

It became clear that no amount of protestation would deter her from insisting that I carried out her wishes. I had no alternative but to accept my situation and resign myself gracefully to the impending ordeal.

A sleepless night was to be expected. I was not disappointed. Over breakfast, Tía said that she was expecting her old friend, the retired Italian consul for coffee, and she would like me to meet him. She said that I would remember such a charming man. I did vaguely remember him, but his name escapes me now. I made my excuses as soon as was decently possible and went to the bedroom. I was completely preoccupied with what I was going to say that evening and the day seemed interminable. I just wanted to get the speech over with!

Lunch was just another interruption and I can't remember what we had, only that I was perplexed. I thought that perhaps Tía would dispense with the ritual tea but no; we had to have the whole works in the garden as usual. Tía was certainly a creature of habit. Soon after tea, we all went to get ready for the banquet. Luckily Helena and I had brought togs for the evening, just in case. I recollected that Tío Manuel and Tía Maruja always used to change for dinner.

Helena came down the marble stairs looking absolutely stunning and the three girls looked gorgeous in their new party dresses. Tía was again in black, but this time she had a beautiful maroon wrap over her shoulders, very elegant. We had arranged to go directly to the Club de Mar, but parking however, proved difficult, and I had to drop the ladies off first and then scamper around for a space some distance away.

As we came into the foyer, we found Mercedes holding forth with a group of people whom she introduced as soon as she caught sight of us. We were given a glass of Cordón Negro whilst we mingled. The *maître d'hôtel* announced that dinner was ready to be served. Much to our surprise, Helena and I were escorted to seats at the top table with the dignitaries. The children sat at a sprig near to us.

It was a most lavish affair and the food was superb. We could see the shimmering Mediterranean through the patio doors, which led to a balcony the whole length of the dining room-cum-ballroom. I was just about getting used to the sound of Spanish being spoken rather rapidly when I heard the sound of a gavel. My heart sank! Mercedes was on her feet. She was a very accomplished speaker; this particular event was a fundraising function as well as the annual ball and money figures quoted sounded very satisfactory. A round of applause was given and then I heard my name being called!

I was up and standing by my chair. I held on to the back of it to steady myself as I surveyed the sea of faces in front of me.

'Señores y señoras, me da muchísma alegría estar de pie en frente de ustedes. Me parece que yo no tengo razón para hablarles en un castellano tan malo, pero se me ha olvidado casi todo. Hace 37 años desde que dejé Almería... Durante aquel tiempo no he hablado ninguna palabra de español...'

I continued in a similar vein, making jokes at my own expense, which they thoroughly enjoyed – I am sure that my Spanish made the jokes sound even more amusing.

Well, it was over. It hadn't been quite as bad as I had feared and the audience had been generous with their applause. We were now on the floor, dancing a *pasadoble*. Every time we passed Mercedes and her husband, Mercedes said, 'Has hecho muy bien. Un montón de gracias.' (It was really good – many thanks.) I was delighted; my biggest worry was not to let Mercedes down. We enjoyed the rest of the evening immensely.

The following morning we went shopping in Almería with Tía. The *paseo* and the side streets leading off it had a marvellous array of shops where you could get just about anything. We sat at one of the cafés on the pavement and had a plate of *churros* and thick hot chocolate. This had to be my favourite. Helena asked Tía where she could get some perfume, as she wanted to dress up and go to the fiesta the next night.

The twins were most anxious to get the correct gear for Mathew, our pet donkey, who we had saved from a terrible life in southern Ireland. Naturally, Tía knew exactly where to get the complete and authentic gear; this was to prove a great success back home when Mathew was used to promote Spanish holidays.

Pura served dinner the next evening quite early so that we could go into Almería in good time to see the start of the fiesta. Tía decided that it was going to be too much jollity for her, and in any case, she had seen it many times before.

We walked down to Cuatro Caminos and caught a bus into town. We got off at Puerta Puchena; that was as far as the bus could go, for the *paseo* was choc-a-block with people. A carnival atmosphere pervaded and we were swept along by the crowd. The girls became separated from us. I was worried, especially when I saw an old woman put her hand in Teresa's pocket. We had given each of them 100 pesetas to spend at the fair.

Rockets, music and drums filled the air and eventually we met up with the girls again. We enquired if Teresa had lost any money. Her face went beetroot red… 'Dad, I dropped the 100-peseta note you gave me and I couldn't bend down to pick it up as the crowd kept pushing me forward. Then an old woman held my arm and put the 100 pesetas in my pocket,' she said. I could not comprehend how anyone so poor would not only resist keeping such a windfall, but would also then go out of her way to restore it to its rightful owner. My immediate instinct was to find the old woman and give her some money, but she had melted away into the crowd. If there is any justice in this world, she should now be in heaven.

It was dawn before we returned home in a taxi. Tía had been worried. What with the oppressive heat and being overtired, we all had a restless night. The next day we took Tía out to dinner as a way of saying thank you for having us, and to give the hard-working Pura a well-earned rest. We went to Tía's favourite restaurant, called Las Palmas, near the Eritaña, but down to a cove on the other side of the road of death, as it was known. It was a lovely place, but expensive… That pleased Tía, who could not wait to tell her sister Soledad and her rich friends that she had been taken to Las Palmas.

Much as we enjoyed staying with Tía, it was time to leave and make our way to Barcelona to see my brothers Luis and Manolo. Next day, we made our fond farewells around mid-morning, and not without a tear or two from the girls and Helena, who had taken a shine to Tía. Even Tía had tears in her eyes! We waved all the way down the drive to the main entrance and stopped a little further on in order to check my map.

We took the N340 road to Tabernas and Sorbas, which passed by the huge cemetery on the outskirts of Almería, where all my ancestors had been buried. I felt that we could not miss the opportunity to see the family vault and to pay my rather belated respects to Tío Manuel, after he had pleaded for me to see him just before he died.

I vaguely remembered where it was, but I did have the vault number, which my sister had written down for me.

Alfuera número 2… Santa Agueda-Dentro número 11, grupo 1, zona 1.

It appeared from the above that the family had other graves outside. These, on inspection, were very old and hard to read. I did not have a key to the vault and had to be cross-examined by an official before he was prepared to let us have an ancient key.

The key fitted a wrought-iron gate to the vault. It felt quite spooky going down the steps to a large chamber, which had three rows of pigeonholes on two sides. Over half were sealed by inscribed slate. I expected to find the vault to be dank and have an obnoxious smell; much to my surprise, it only smelt a bit stuffy.

Walking past the rows of niches, I recognised many names – the newest being that of Tío Manuel. I had taken a colour photograph of my grandparents' oil paintings hanging above the oak panelling in the dining hall and was now also to see their burial place. I found them both next to each other, on a row at about ground level. La Marquesa Ana Echeverría Patrullo, Murió 2-7-1919, 62 años de edad; D. Guillermo López Rull, Murió 3-7-1929, 77 años de edad. My father came next: Guillermo López Echeverría, Murió 9-2-1928, 46 años de edad.

I was so glad that the three girls had this opportunity to see the roots, as you might say, of their parental ancestors.

We had spent a long time at the cemetery so we really had to get a move on. The road up to where some of the western films were being made, now called Mini Hollywood, was quite good. After that, the surface and bends through the Sierra Alhamilla made driving difficult. But in spite of the road, we made good progress. I was obviously becoming more used to things and was pleasantly surprised when the going got better on the odd occasion.

We arrived at the outskirts of Sorbas and immediately were struck by the spectacular appearance of the sparkling white houses, virtually built on top of each other on the edge of the cliffs, as we entered the town. Although tempted to stop, we decided to press on to Vera, whilst the light was still good. Unfortunately, the road got steadily worse and we were also becoming worried at seeing so few cars. Now only the odd *camión* passed by, going the other way. The sheer drop as the road hugged the side of the mountains was more than enough to make anyone nervous.

We had travelled for a considerable period without seeing so much as a lorry, when coming round a wicked bend, we saw a massive *camión*, partly hanging over the edge of the road. To our horror, we saw a man lying underneath. It was a job to stop and park anywhere safely, as the road was narrow and the next bend was but a few metres away. We took a chance and I got out of the car and peered under the lorry, saying in a very unsure voice, '*¿Está bien? ¿Puedo ayudarle?*' To my relief, the man opened his eyes and said that he had been waiting over two hours for a rescue vehicle. We offered to pass on a message at the next garage… Perhaps there was one at Vera? *'No, gracias, creo que ayuda ya venga,'* he said. We gave him a drink of Fanta and we went on our way.

We decided to turn off the main road and make for Garrucha on the coast and hopefully find somewhere to sleep for the night. We were lucky and found a *fonda* which looked clean and adequate. It was here that the Americans lost an atomic bomb a few years later and it took them several months to find it again!

We got into our stride early next morning and made for Vera. Then instead of going on to Huercal, we decided to keep to the coast road. This proved a bit of a disaster, as the road was not only terrible, but also very dangerous and busy. Eventually, we got to Cartagena in the dark and decided that we could not go further as oncoming traffic did not like our headlights and kept flashing us. We had a look round for a hotel, but at that time of the night it proved a fruitless task. We pulled off the main road and went down a track to the beach where some people were still bathing. After a while, we decided that we were too conspicuous to stay on the beach all night and drove back to the main road and parked under a bridge, close to one of its supports.

We had something to eat and drink and settled down for the night. Being tired we soon went to sleep, but not for long! A torch shone in our faces and we found a couple of Civil Guards looking rather curiously at us. After explaining our predicament, they said they would keep an eye on us and see that we were safe. Keeping an eye on us meant that every two hours, torches were flashed on our faces. In the end, we gave up trying to sleep and offered them a drink and biscuits. At dawn, the girls wanted a swim, after which we wandered a little way into town and bought more drinks and nibbles to keep us going on the journey.

We had a brief stay in Alicante and then kept to the coast road to Valencia, where we decided to stop for the night and have one of the famous paellas. After all, it's the home of paella! I regret to say that to this day, we have not had a worse paella, and yet we dined at a good quality restaurant.

The hotel had no air conditioning and, consequently, we slept with the bedroom windows wide open. Next morning you would have been forgiven if you had thought that we had caught a dreadful disease by the look of our faces, legs and arms – they were covered in mosquito bites! They must have been a very virulent variety, which were no doubt bred nearby in the rice paddy fields.

We had about another 150 miles to go for Barcelona, where Luis had booked a hotel for us near to his apartment. Although the road was good, the traffic was extremely heavy. Nevertheless, we were determined to get to Barcelona by mid-afternoon and fortunately the rest of the journey went well.

The approaches to Barcelona were intimidating. I think I counted seven traffic lanes, but luckily the hotel proved relatively easy to find. We booked ourselves in at about 3 o'clock in the afternoon. A good soak was at the top of the agenda; next was to put on some glad rags and then find Luis's place at no. 17 Ricardo Calvo. To our amazement we found it in minutes as it was very close to the hotel.

My old ticker was beginning to race with excitement… I still had my memories of leaving Luis in Málaga, 37 years earlier. We got into the lift and alighted on the second floor… This was it. Ring the bell… But they must have seen us, as the door opened before I got my finger on the bell.

'¡Hola Emilio!' said Luis, with a smile, which reached from ear to ear. I stuttered, 'I am so glad to see you,' in English. Luis looked puzzled; I had always imagined that he would speak English – but then again, why would he?

María was standing behind, but came forward and we all got a bear hug and a great smile. I had to do my stuff and start speaking in Spanish. We entered the spacious and airy apartment and, in spite of the language barrier, we seemed to be able to carry on several conversations simultaneously. Maite, Luisa and Guillermo came into the room – more embraces – but of course, we knew Luisa. She had stayed with us in England and her English was good so this was to take some of the pressure off me, thank goodness!

I found myself staring at Luis. He was so different to what I had visualised. He appeared so much older than me, and yet I knew that he could only be about three years older. María and her two daughters then began to bring a great selection of tapas and corks began to pop!

There was much to talk about and yet we knew there was little time. Soon we were being shepherded along to a nearby roof-top restaurant overlooking the city. It was a delightful place and vines grew in and out of the trellis above the shallow walls. Bunches of grapes and electric lights hung down from the wooden beams above.

Luis had reserved the whole of the roof garden restaurant. We sat around a large table made up from several smaller tables put together. We had only just got going again with the conversations that we had left at the apartment, when Luis shouted in Spanish, 'Well, look who's here!'

And there was Manolo, my other brother, whom you could have taken as my twin. He was with his wife, whom he introduced to Helena and me as Lola.

Manolo and I got into a huddle standing near the table, whilst Lola sat down between Helena and María. Manolo, I learnt, was the captain of one of the police precincts in Barcelona. When I left Almería, Manolo was only a baby and was living with my mother, so I did not know him. All I knew was that he existed and that there appeared to be some doubt as to his parentage. But after seeing him now, I felt sure that he was a true blood brother.

The repast went on into the early hours of the morning, the proceedings being punctuated by the odd speech of welcome. Luis began to tell me about his experiences after escaping from the school in Málaga, but we had to leave it for another day, as the children were by now getting well past their sell-by dates!

Next morning, we went into the centre of Barcelona and had a good look around the wonderful shops, and apart from the many boutiques, we liked the department store called El Corte Inglés.

We had a late lunch, which María had prepared whilst we were sightseeing. Over lunch, Luis said that I must visit my mother who was now living in Barcelona and was expecting to see me. It was agreed that I would pick him up in the morning so that he could show me the way.

When we had finished lunch, Luis and I went on chatting in his study. Our girls got together with Luis's children and Helena vanished into the main bedroom with María. Half an hour or so later, Helena came rushing into the study saying that we had to go, there and then.

'What on earth is the matter?' I said.

'Did you know that your brother was an officer in the Wehrmacht?' Helena said in a most agitated manner. I replied that I didn't know.

'There is a photograph in the bedroom which shows him wearing an Iron Cross, would you believe it?' she said.

I was somewhat stunned by the news and did not readily offer any explanation. I emphasised that it had been a long time ago.

Luis all this time sat in his armchair, with eyes staring at Helena and I am sure, wondering what on earth had possessed her.

'Have you stopped to think, angel, that there might have been a perfectly good reason why he joined the German Army?'

'Do you realise you were on opposite sides?' she retorted.

At this point, Luis got up and, standing in the middle of the room, he said, 'I think that I owe you an explanation' (in Spanish). He carried on with what was obviously going to be a long story, which I would have to translate for Helena's sake. Luis continued:

'The fact is that I hated the Communists. After you left me at our boarding school, Emilio, a bunch of Communists and anarchists burst into the school and shot all the masters that they could find. They dragged their bodies out into the main street. They also murdered a couple of the older boys too. Luckily for me, I was in the dormitory when I heard the gunfire and I hid up in the roof, for they were searching around for other masters and older boys.

'When the coast was clear, I ran to the port, ducking and weaving in order to avoid the murderous mobs that were roving the streets for likely victims. An old cargo boat was moored at the quay. I was determined to get to Almería, but I wasn't sure where this boat was going as the name had been painted out. I was about to jump on board when I noticed that a group of cut-throats were doing a search. They brought out a couple of very frightened men and left the ship. A member of the crew jumped onto the quay and took the fore and aft cables off the bollards and climbed back on board. At that moment, I took a chance and scrambled onto the deck and hid behind some deck cargo.

'I had no idea where I was going and I finally ended up at

Cartagena, having passed Almería. I had no choice but to walk all the way back to Almería, as it was far too dangerous to try to get a lift. I lived by my wits for several months, as I was unable to make contact with the family. The ones who had not been shot or arrested were being watched. I made my way up the coast and ended up as a farm labourer until the end of the Civil War, when I moved to Barcelona.

'Although Franco had won the war, things in Spain were very bad, so when the Blue Division was formed as a volunteer force to support the Germans in their fight against the Russians, I joined up, for the dreadful scenes I had witnessed in Málaga were still as vivid in my mind as if they had happened only yesterday. I wanted to avenge the horrors they had perpetrated; I owed it to those innocent masters who had been mown down in front of my eyes.

'After a long period of training in the winter of 1941, the Blue Division moved up to the front. The Division fought alongside the German Army as far as the encirclement of Leningrad in September 1941, where we dug in through the winter and fought some of the most fierce battles of the war. I was taken prisoner at one stage, but I managed to escape and rejoin my unit. I was shot in the chest... The bullet went right through me and passed very near to my heart.'

'Can we have a look?' said one of the girls and he pulled his shirt open to reveal a pinky brown round mark on his chest, just below his shoulder.

'This finger,' he indicated a gap on his left hand, 'I lost in hand-to-hand fighting at Schusselburg in September 1941, when we completed the encirclement of Leningrad. The Blue Division had suffered huge casualties. They left over 5000 buried in the snows of Gringorowo, Podvnedje, Osselek and at the Winter Palace of the Czars in St Petersburg.

'In October 1943, the Blue Division received orders to withdraw from the front line and return to Spain. It was then that the Wehrmacht authorised the inclusion of 2133 Spanish volunteers from the Blue Division. These soldiers rallied behind Colonel García Navarro, Lieutenant López de la Torre and myself, and chose to continue the fight. Strangely enough, the names of the colonel and the lieutenant were similar to the names of two branches of our family, but we never found out if there was any connection.

'I was one of them. We were attached to the fourteenth Antitank Core, under the name of Legión Azul, but I then wore the uniform of a lieutenant in the German Army, hence the photograph showing the Iron Cross, given to me personally by General Lindeman. My other decorations are Spanish military medals and to finish, I might as well tell you that I am, to this day, the Secretary of both the División Azul and the Legión Azul Branch of Cataluña.'

With this, Luis looked at each of us in turn and sat down.

Without a shadow of a doubt, Helena had come on board. She got up to say sorry that she had been so angry and explained that she understood that what he did, he had to do. Both gave each other a hug and everything was forgiven.

After the short history delivered to us by Luis, Manolo and I stayed behind in Luis's study, where we three all did some catching up. Manolo seemed very reserved but he had an unmistakable presence, probably with him being a policeman. However, I do believe that his wife Lola wore the trousers!

I was intrigued to find out what happened to Luis when the Allies won the war. The Legión Azul, or what was left of it, was ordered to abandon their camp at Konigsberg, where their colours were paraded for the last time in front of the general and chief of staff. They marched back to Spain under Colonel García Navarro. Luis once again remained behind, as his sense of duty prevented him from deserting his German allies at such a crucial moment. He surrendered with his unit to the Americans, whom he said were most magnanimous... For him, it turned out to be the best part of the war! The American high command saw that the prisoners were treated with respect and were well fed. Luis said that the majority who retreated from the Russian front were undernourished and very thin.

★

It was getting late and we had to return to the hotel. It had been a most eventful day, but right now we were all ready for bed. Next day, I picked up Luis as arranged and made for a nursing home where my mother was staying. I was apprehensive about the meeting after so many years, particularly as she had not been too pleased when I left Spain on my own.

Luis's presence was a great help. Our mother was at least pleased to see him. My brother sort of introduced me to my own mother! This spoke volumes... Neither managed to rekindle the embers. We just looked at each other in a curiously detached way. Why did we bother? It was a hopeless case. Luis and I left after a very short while as conversation was forced and painful and my mother, I felt, was glad to say goodbye. Both of us knew we would never see each other again.

The girls and I had to pack in some sightseeing, as we would be leaving in a couple of days' time. The next day we were given a whirlwind tour by Luis and his children. Guillermo was great fun and our three girls got very fond of him.

First we went to Monserrat, a monastery by the sea, high up on the cliffs overlooking Barcelona. Helena did not like going up the Tibidabo on a funicular railway which went up the side of the mountain. It was a spectacular view. Next we went to El Pueblo Español, which we all enjoyed. It represented each region of Spain by way of architecture, food and vegetation, as well as artefact products typical of each area. We were pretty exhausted by now, and as María was going to do another one of her mammoth spreads for dinner as a going-away celebration, we decided to call it a day at teatime. We returned to the hotel for a nice long shower and a couple of drinks out of the mini-bar.

The repast surpassed anything that María had done before and champagne was much in evidence. It would be a night to conjure up when back in England. Everyone had been so kind and generous. But there we were, saying our affectionate goodbyes time and time again. Eventually, we did take our leave in the very early hours of the morning. We had to walk to the hotel as I had had far too much to drink to be able to drive. We left the car parked outside Luis's apartment.

Feeling somewhat delicate, we vacated our rooms the next morning, paid the bill and left our cases in the hallway in order to pick them up with the car.

When we arrived at the high-rise block, we found Luis and his family looking out for us from their windows. When we walked into the apartment, we found lots of small parcels awaiting us. We were all handed one or more. They were mementos of our visit. We were about to go when Helena dashed off with María; I had thought that something had been worrying her and that she seemed stressed, but I put it down to the emotions which were welling up inside us all.

Presently, Guillermo came rushing into the sitting room holding up what looked suspiciously like Helena's engagement ring! Helena must have heard Guillermo and came rushing into the room… She grabbed Guillermo and gave him the biggest kiss of his life. It would appear that Helena had noticed that her engagement ring had come off her finger the previous night and she had hardly slept a wink. She couldn't have got to Luis's apartment quick enough that morning in the hopes that it might have been found.

When asked where he found it, Guillermo said, 'Well, you will never believe it.' He dragged Helena outside into the main street and pointed to the left-hand overrider on the rear bumper of our car. 'Here,' he said. 'It was resting on top and the sun was shining on it, making the diamonds sparkle. Otherwise I would never have seen it.'

This had to be a miracle… The car had been parked in the main street all night.

We got in the car together with Luis, Guillermo and Mateas… I am not quite sure how they all fitted in, but somehow we picked the suitcases up and sped away to the airport, where I handed the car back to a BEA representative and apologised for the horrid dent in its side. Fortunately, that was covered by insurance.

We checked in right away as we were late and in many ways, I was glad to have to rush, as another prolonged farewell might have proved too much for us. But as we looked back to wave, I noticed that Luis had tears in his eyes… So he was a softy after all!

Well, my ambition had been achieved. After looking forward to this visit, for 37 years, I had no desires to stay and it was back to dear old England, my new home, where I had found happiness.